Nina Nowak

Supermassive black holes in peculiar galaxies

Nina Nowak

Supermassive black holes in peculiar galaxies

Exploring the relations between the central black hole mass and host galaxy properties in low-mass galaxies, merger remnants and pseudobulges

Südwestdeutscher Verlag für Hochschulschriften

Impressum/Imprint (nur für Deutschland/ only for Germany)
Bibliografische Information der Deutschen Nationalbibliothek: Die Deutsche Nationalbibliothek verzeichnet diese Publikation in der Deutschen Nationalbibliografie; detaillierte bibliografische Daten sind im Internet über http://dnb.d-nb.de abrufbar.
 Alle in diesem Buch genannten Marken und Produktnamen unterliegen warenzeichen-, marken- oder patentrechtlichem Schutz bzw. sind Warenzeichen oder eingetragene Warenzeichen der jeweiligen Inhaber. Die Wiedergabe von Marken, Produktnamen, Gebrauchsnamen, Handelsnamen, Warenbezeichnungen u.s.w. in diesem Werk berechtigt auch ohne besondere Kennzeichnung nicht zu der Annahme, dass solche Namen im Sinne der Warenzeichen- und Markenschutzgesetzgebung als frei zu betrachten wären und daher von jedermann benutzt werden dürften.

Verlag: Südwestdeutscher Verlag für Hochschulschriften Aktiengesellschaft & Co. KG
Dudweiler Landstr. 99, 66123 Saarbrücken, Deutschland
Telefon +49 681 37 20 271-1, Telefax +49 681 37 20 271-0
Email: info@svh-verlag.de
Zugl.: München, Ludwig-Maximilians-Universität, Diss., 2009

Herstellung in Deutschland:
Schaltungsdienst Lange o.H.G., Berlin
Books on Demand GmbH, Norderstedt
Reha GmbH, Saarbrücken
Amazon Distribution GmbH, Leipzig
ISBN: 978-3-8381-1503-0

Imprint (only for USA, GB)
Bibliographic information published by the Deutsche Nationalbibliothek: The Deutsche Nationalbibliothek lists this publication in the Deutsche Nationalbibliografie; detailed bibliographic data are available in the Internet at http://dnb.d-nb.de.
 Any brand names and product names mentioned in this book are subject to trademark, brand or patent protection and are trademarks or registered trademarks of their respective holders. The use of brand names, product names, common names, trade names, product descriptions etc. even without a particular marking in this works is in no way to be construed to mean that such names may be regarded as unrestricted in respect of trademark and brand protection legislation and could thus be used by anyone.

Publisher: Südwestdeutscher Verlag für Hochschulschriften Aktiengesellschaft & Co. KG
Dudweiler Landstr. 99, 66123 Saarbrücken, Germany
Phone +49 681 37 20 271-1, Fax +49 681 37 20 271-0
Email: info@svh-verlag.de

Printed in the U.S.A.
Printed in the U.K. by (see last page)
ISBN: 978-3-8381-1503-0

Copyright © 2010 by the author and Südwestdeutscher Verlag für Hochschulschriften Aktiengesellschaft & Co. KG and licensors
All rights reserved. Saarbrücken 2010

Zusammenfassung

In dieser Arbeit wird über die Messung der Masse von supermassiven schwarzen Löchern im Zentrum von vier Galaxien mit der Methode der stellaren Dynamik berichtet. Grundlage hierfür sind Beobachtungen im nahen Infrarot (1.9 – 2.5 μm), die mit dem Feldspektrographen SINFONI am Very Large Telescope (VLT) durchgeführt wurden. Anhand dieser Daten konnten die Bewegungen der Sterne im Einflußbereich des schwarzen Lochs gemessen und dessen Masse mit Modellrechnungen rekonstruiert werden.

Derartige Messungen konnten bisher hauptsächlich mit dem Hubble Space Telescope im optischen Spektralbereich in massiven, staubfreien Galaxien durchgeführt werden. Mit SINFONI und der Technik der adaptiven Optik ist es jetzt möglich, die Masse von schwarzen Löchern in staubigen, massearmen und leuchtschwachen Galaxien zu messen. Dies ist notwendig, um die Zusammenhänge zwischen der Entwicklung von Galaxien und dem Wachstum von schwarzen Löchern verstehen zu lernen.

Die leuchtschwache elliptische Galaxie NGC 4486a hat eine Staubscheibe im Zentrum und ist mit einer Geschwindigkeitsdispersion von 110 km s^{-1} im unteren Bereich des Massenspektrums von Galaxien angesiedelt. Mit der sogenannten "Maximum Penalized Likelihood" Methode wurde die Geschwindigkeitsverteilung der Sterne gemessen. Die Masse des schwarzen Lochs wurde mit Hilfe der Schwarzschild-Methode zu $M_\bullet = (1.26 \pm 0.41) \times 10^7$ M$_\odot$ (68% Vertrauensintervall) bestimmt. Dies stimmt mit den Vorhersagen der Beziehung zwischen $M_\bullet$ und Geschwindigkeitsdispersion σ der Sterne im zentralen Bereich (dem "Bulge") der Galaxie ($M_\bullet$-σ-Relation) überein, und auch mit der Relation zwischen $M_\bullet$ und Bulgemasse ($M_\bullet$-M_{bulge}-Relation), die beide auf den Messungen in massereichen elliptischen Galaxien basieren.

Wenn Galaxien ähnlicher Masse kollidieren und miteinander verschmelzen, so bildet sich eine neue Galaxie. Die schwarzen Löcher sinken ins Zentrum und verschmelzen dort, und durch Transport von Gas ins Zentrum kann nukleare Aktivität und somit das weitere Wachstum des schwarzen Lochs ausgelöst werden. Fornax A, eine Radiogalaxie im Fornax-Galaxienhaufen, ist das Resultat einer solchen Verschmelzung. Die Geschwindigkeitsverteilung und die Masse des zentralen schwarzen Lochs wurden auf dieselbe Weise gemessen wie in NGC 4486a. Mit $M_\bullet = 1.5^{+0.75}_{-0.8} \times 10^8$ M$_\odot$ (99.7% Vertrauensintervall) folgt auch Fornax A der $M_\bullet$-σ-Relation. Jedoch ist $M_\bullet$ ungefähr viermal kleiner, als man es von der Masse des Bulges erwarten würde.

Ein Bulge kann unterschiedliche Erscheinungsformen haben, abhängig von seinem Entstehungsmechanismus. Es wird angenommen, dass ein klassischer Bulge bei der Kollision von Galaxien entsteht und in etwa die Eigenschaften einer elliptischen Galaxie aufweist. In einer Scheibengalaxie gibt es jedoch Mechanismen, die Gas vom äußeren Teil der Scheibe ins Zentrum transportieren,

Zusammenfassung

wo dann Sterne und somit ein Bulge gebildet werden. Ein Bulge, der auf diese Weise entsteht, hat in etwa die Eigenschaften einer Spiralgalaxie und man nennt ihn "Pseudobulge". In den Galaxien NGC 3368 und NGC 3489, die beide sowohl einen Pseudobulge, als auch einen kleinen klassischen Bulge besitzen, wurde die Masse der zentralen schwarzen Löcher zu $M_\bullet = 7.5 \times 10^6$ $M_\odot$ mit einem RMS-Fehler von 1.1×10^6 $M_\odot$ (NGC 3368) und $M_\bullet = (6.00^{+0.56}_{-0.54}|_{\text{stat}} \pm 0.64|_{\text{sys}}) \times 10^6$ $M_\odot$ (68% Vertrauensintervall, NGC 3489) bestimmt. Beide stimmen mit den Vorhersagen der $M_\bullet$-σ-Relation überein. Die $M_\bullet$-M_{bulge}-Relation jedoch sagt vielfach größere Massen vorher, geht man von den Gesamtmassen der Bulges (Pseudobulge + klassischer Bulge) aus. Die Masse der klassischen Bulge-Komponente scheint hier ein besserer Indikator für die Masse des schwarzen Lochs zu sein.

Abstract

This thesis reports on the measurement of the masses of supermassive black holes in the centres of four galaxies using stellar dynamics. It is based on observations in the near-infrared (1.9 − 2.5 μm) with the integral-field spectrograph SINFONI at the Very Large Telescope (VLT). These data were used to determine the motions of stars in the very centre of the galaxies, which then were modelled to derive the mass of the central black hole.

Such measurements were until now restricted to massive and dust-free galaxies, observed with the Hubble Space Telescope in the optical wavelength range. With SINFONI and the technique of adaptive optics it is now possible to measure the mass of supermassive black holes also in dusty, low-mass and faint galaxies. This is essential in order to understand the correlation between the evolution of galaxies and the growth of central black holes.

The faint elliptical galaxy NGC 4486a contains a nuclear disc of dust and stars. With a velocity dispersion of 110 km s^{-1} it belongs to the class of low-mass galaxies. With the so-called "Maximum Penalized Likelihood" method the velocity distribution of the stars was measured. The mass of the central black hole, determined using the Schwarzschild method, is $M_{\bullet} = (1.26 \pm 0.41) \times 10^7$ M$_{\odot}$ (68% confidence limit). This is in agreement with the relation between $M_{\bullet}$ and velocity dispersion σ of the galaxy bulge ($M_{\bullet}$-σ relation), as well as with the relation between $M_{\bullet}$ and bulge mass ($M_{\bullet}$-M_{bulge} relation), both of which are based on measurements of massive elliptical galaxies and bulges.

When galaxies of similar mass merge, a new galaxy forms. The central black hole of each galaxy sinks to the centre where they eventually coalesce. If the progenitor galaxies are gas-rich, gas might be transported to the centre where it might trigger nuclear activity and thus the further growth of the central black hole. Fornax A, a radio galaxy located in the outskirts of the Fornax galaxy cluster, experienced a recent major merger. Its velocity distribution and the mass of the central black hole were measured in the same way as for NGC 4486a. With $M_{\bullet} = 1.5^{+0.75}_{-0.8} \times 10^8$ M$_{\odot}$ (99.7% confidence interval) also Fornax A follows the $M_{\bullet}$-σ relation, but $M_{\bullet}$ is a factor of about four smaller than expected from the $M_{\bullet}$-M_{bulge} relation.

Bulges appear in different forms, depending on their formation mechanism. It is believed that classical bulges form via mergers and have the general properties of elliptical galaxies. Pseudobulges are thought to be the result of secular evolution, where gas is transported from the outer parts of a spiral galaxy to the centre, where it forms stars and builds a bulge with properties similar to the outer disc. The galaxies NGC 3368 and NGC 3489 both host a pseudobulge and a small classical bulge component. The derived masses of the central black holes are $M_{\bullet} = 7.5 \times 10^6$ M$_{\odot}$ with an RMS error of 1.1×10^6 M$_{\odot}$ for NGC 3368 and $M_{\bullet} = (6.00^{+0.56}_{-0.54}|_{\text{stat}} \pm 0.64|_{\text{sys}}) \times 10^6$ M$_{\odot}$ (68% confidence limit) for NGC 3489. Both masses are in agreement with the $M_{\bullet}$-σ relation. The $M_{\bullet}$-

Zusammenfassung

M_{bulge} relation, however, predicts several times larger black hole masses when the total bulge mass (including pseudobulge and classical bulge) is considered. The mass of the small classical bulge component in this case seems to be a better indicator of the black hole mass.

Contents

List of figures	ix
List of tables	xiii

1 Black holes in galaxies   1
- 1.1 A brief history of black holes 1
- 1.2 The co-evolution of black holes and their hosts 4
- 1.3 Black hole mass measurements – status today 7
  - 1.3.1 Measurement techniques 7
  - 1.3.2 Catalogue of $M_\bullet$ measurements 11
- 1.4 Limitations of the $M_\bullet$-σ and the $M_\bullet$-L_{bulge} relation 12
  - 1.4.1 Stars vs. gas 17
  - 1.4.2 Pseudobulges vs. classical bulges 17
  - 1.4.3 Barred vs. unbarred galaxies 23
  - 1.4.4 Spiral vs. elliptical galaxies 24
  - 1.4.5 Low-σ and bulgeless galaxies 24
  - 1.4.6 High-σ and core galaxies 25
  - 1.4.7 AGN vs. inactive galaxies 26
- 1.5 Outline of the thesis 31

2 Observations and data reduction   33
- 2.1 SINFONI 34
  - 2.1.1 Adaptive optics 34
  - 2.1.2 Laser guide star 37
- 2.2 Selection of galaxies 39
- 2.3 Observations 41
- 2.4 Data reduction 51

Contents

3 Stellar kinematics   **59**
- 3.1 Extraction techniques 59
- 3.2 Spectral features 64
- 3.3 Stellar templates 66
- 3.4 Performance of FCQ and MPL in the near-IR 68
  - 3.4.1 Fourier Correlation Quotient method (FCQ) 70
  - 3.4.2 Maximum Penalized Likelihood method (MPL) 74
- 3.5 Recipe to obtain LOSVDs from K-band SINFONI spectra 83

4 The supermassive black hole of NGC 4486a   **89**
- 4.1 Introduction 90
- 4.2 Observations and data reduction 91
- 4.3 Kinematics 93
- 4.4 Imaging 96
- 4.5 Schwarzschild modelling 99
- 4.6 Results 100

5 The supermassive black hole of Fornax A   **105**
- 5.1 Introduction 106
- 5.2 Data and data reduction 107
  - 5.2.1 SINFONI data 107
  - 5.2.2 Longslit data 111
  - 5.2.3 Imaging 111
- 5.3 Stellar kinematics 114
  - 5.3.1 Initial parameters 115
  - 5.3.2 Kinematic template stars 115
  - 5.3.3 Error estimation 116
  - 5.3.4 The kinematics of Fornax A 116
- 5.4 Line indices 126
- 5.5 Dynamical models 130
  - 5.5.1 The stellar dynamical K_s-band mass-to-light ratio Υ 131
  - 5.5.2 The black hole mass $M_\bullet$ 133
- 5.6 Summary and discussion 141

6 The pseudobulge galaxies NGC 3368 and NGC 3489   **145**
- 6.1 Introduction 146

6.2	Imaging	149
	6.2.1 Imaging data and calibrations	149
	6.2.2 NGC 3368	150
	6.2.3 NGC 3489	156
6.3	Spectroscopy	157
	6.3.1 Data and data reduction	157
	6.3.2 Stellar kinematics in NGC 3368	166
	6.3.3 Gas kinematics in NGC 3368	170
	6.3.4 Line strength indices for NGC 3368	170
	6.3.5 Stellar kinematics in NGC 3489	174
	6.3.6 Line strength indices for NGC 3489	174
6.4	Dynamical modelling of NGC 3368	176
	6.4.1 Construction of the stellar luminosity profile	176
	6.4.2 Dynamical models	178
	6.4.3 Results	179
	6.4.4 Discussion	184
6.5	Dynamical modelling of NGC 3489	186
	6.5.1 Construction of the luminosity profile for modelling	186
	6.5.2 Dynamical models	188
	6.5.3 Results	188
	6.5.4 Discussion	194
6.6	Summary and discussion	194

7 Concluding remarks   **201**

Bibliography   **211**

Acknowledgements   **235**

List of figures

1.1	$M_\bullet$-$M_{B,\text{bulge}}$ relation	3
1.2	$M_\bullet$-σ relation	4
1.3	$M_\bullet$-σ relation for different galaxy samples	18
1.4	$M_\bullet$-L_K relation for different galaxy samples	27
2.1	Inside view of SPIFFI	35
2.2	Principle of integral-field spectroscopy	36
2.3	Sphere of influence as a function of σ and D	40
2.4	Spatial dithering	45
2.5	Atmospheric transmission	54
2.6	3D datacube	55
3.1	Change of LOSVD shape with h_3 and h_4	62
3.2	Stellar kinematic template stars	69
3.3	Template spectrum convolved with Gaussian LOSVDs	71
3.4	Simulations with FCQ	73
3.5	Kinematic parameters for different S/N	76
3.6	Kinematic parameters for different σ	77
3.7	Kinematic parameters at low σ_{in} and large Δv	79
3.8	Kinematic parameters at low σ_{in} using parametric fits	79
3.9	Kinematic parameters for different h_3 and h_4	81
3.10	Kinematic parameters for different wavelength regions	82
3.11	Fit of template to galaxy	83
3.12	Smoothing as a function of S/N and σ	85
3.13	LOSVD change with α	86
4.1	SINFONI image of NGC 4486a	92

List of figures

4.2	PSF of NGC 4486a	93
4.3	Stellar kinematics of NGC 4486a	95
4.4	Major axis kinematics of NGC 4486a	97
4.5	Surface brightness profile of NGC 4486a	98
4.6	Dynamical models for NGC 4486a	101
4.7	χ^2 difference between the best models with and without black hole	102
5.1	SINFONI images of Fornax A	108
5.2	PSF of Fornax A	110
5.3	SOFI and *HST* WFPC2 images of Fornax A	112
5.4	Surface brightness profile of Fornax A	113
5.5	Spectrum and LOSVD of Fornax A	117
5.6	Stellar kinematics of Fornax A	119
5.7	Convolved 25mas kinematics of Fornax A	120
5.8	Near-IR line indices of Fornax A	124
5.9	H_2 emission of Fornax A	125
5.10	Kinematic parameters as a function of [Ca VIII] contribution	127
5.11	Fit to broadened template spectra with different [Ca VIII] contributions	127
5.12	[Ca VIII] contribution to the SINFONI spectra of Fornax A	128
5.13	χ^2 as a function of Υ for longslit-only models of Fornax A	132
5.14	$\Delta\chi^2$ as a function of $M_\bullet$ and Υ for Fornax A 100mas data	132
5.15	Dynamical models for Fornax A	135
5.16	Dynamical models for the folded Fornax A data	137
5.17	Model fits to the kinematics of Fornax A	138
5.18	χ^2 difference between the best models with and without black hole	140
5.19	Anisotropy as a function of radius of Fornax A	141
6.1	Global bulge-disc decomposition of NGC 3368	151
6.2	Local V_{dp}/σ estimates of NGC 3368	152
6.3	Isophotal ellipse fits for NGC 3368	153
6.4	Isophotal maps of NGC 3368	154
6.5	Decomposition of the photometric bulge of NGC 3368	155
6.6	Isophotal maps of NGC 3489	158
6.7	Isophotal ellipse fits for NGC 3489	159
6.8	Global bulge-disc decomposition of NGC 3489	160
6.9	Local V_{dp}/σ estimates of NGC 3489	161

List of figures

6.10 Decomposition of the photometric bulge of NGC 3489 162
6.11 SINFONI PSF of NGC 3368 164
6.12 SINFONI PSF of NGC 3489 164
6.13 SINFONI images of NGC 3368 and NGC 3489 165
6.14 Stellar kinematics of NGC 3368 167
6.15 *HST* WFPC2 $B - I$ colour map of NGC 3368 168
6.16 H_2 gas emission and velocity of NGC 3368 171
6.17 Near-IR line indices of NGC 3368 and NGC 3489 173
6.18 Stellar kinematics of NGC 3489 175
6.19 Dynamical models for NGC 3368 180
6.20 χ^2 as a function of $M_\bullet$, Υ_{bulge} and Υ_{disc} for NGC 3368 181
6.21 Model fits to the kinematics of NGC 3368 182
6.22 χ^2 difference between the best models with and without black hole 183
6.23 Enclosed mass as a function of radius in NGC 3368 187
6.24 Dynamical models with SAURON and OASIS data for NGC 3489 189
6.25 Dynamical models for NGC 3489 191
6.26 Model fits to the kinematics of NGC 3489 192
6.27 $\Delta\chi^2$ as a function of $M_\bullet$, Υ_{bulge} and Υ_{disc} for NGC 3489 193
6.28 NGC 3368 and NGC 3489 in the $M_\bullet$-σ and $M_\bullet$-M_K relation 199

7.1 Stellar kinematics of NGC 5102 202
7.2 Updated $M_\bullet$-σ relation 203
7.3 Updated $M_\bullet$-L_K relation 205

List of tables

1.1	Galaxies with reliable black hole mass measurements	13
2.1	Observed galaxies and their properties	42
2.2	Observation log	46
3.1	Most important spectral absorption features in the K-band	65
3.2	Stellar kinematic template stars	67
3.3	CO equivalent widths	70
4.1	$M_\bullet$ and mass-to-light ratios of NGC 4486a	100
5.1	CO equivalent widths of the stellar kinematic template stars	116
5.2	Mean near-IR line indices of Fornax A	129
5.3	$M_\bullet$ and Υ with 3 σ errors of Fornax A	139
6.1	Properties of NGC 3368 and NGC 3489	147
6.2	H_2 emission line properties of NGC 3368	171
6.3	Near-IR line strength indices of NGC 3368 and NGC 3489	172
6.4	$M_\bullet$ and mass-to-light ratios of NGC 3368	184
6.5	$M_\bullet$ and mass-to-light ratios of NGC 3489	195
6.6	$M_\bullet$ and mass-to-light ratios for the folded data of NGC 3489	195
7.1	Black hole masses measured in Chapters 4-6	203

1

Black holes in galaxies

1.1 A brief history of black holes

The existence of black holes was first postulated in 1783 by the geologist and astronomer John Michell. Based on Newton's law of gravity he suggested there might be objects so massive that not even light could escape (Michell, 1784):

> *If the semi-diameter of a sphaere of the same density with the sun were to exceed that of the sun in the proportion of 500 to 1, a body falling from an infinite height towards it, would have acquired at its surface a greater velocity than that of light, and consequently, supposing light to be attracted by the same force in proportion to its vis inertiae, with other bodies, all light emitted from such a body would be made to return towards it, by its own proper gravity.*

He proposed that the detection of such bodies could be possible via the observation of motions of luminous objects around them:

> *Yet, if any other luminous bodies should happen to revolve about them we might still perhaps from the motions of these revolving bodies infer the existence of the central ones with some*

1.1. A BRIEF HISTORY OF BLACK HOLES

degree of probability, as this might afford a clue to some of the apparent irregularities of the revolving bodies, which would not be easily explicable on any other hypothesis.

A few years later, in 1796, the mathematician Laplace published basically the same idea (Laplace, 1796). At these times nobody really believed that such objects could possibly exist, thus this idea fell into oblivion for many years. In 1916, after Einstein published his theory of general relativity (Einstein, 1915), the basis of further black hole studies, Schwarzschild (1916) succeeded to find an exact solution to Einstein's field equations. The Schwarzschild solution describes the gravitational field outside a spherically symmetric, non-rotating mass and defines the gravitational radius (the "Schwarzschild radius") of a black hole as the radius where the Schwarzschild metric becomes singular. It is the same radius Laplace found by using Newton's laws and letting the escape velocity be the speed of light.

In the following decades, some work was done on the theoretical side, e.g. by Chandrasekhar (1931), whose work on white dwarfs led to an understanding of mass limits, which determine whether a star ends its life as a white dwarf, a neutron star or a black hole. Kerr (1963) found a solution to Einstein's field equations that described rotating black holes. The term "black hole" was coined in 1968 by John Wheeler during a talk.

The interest in black holes then shifted from theory more to the observational side, as the proof that black holes really exist was still missing. The first convincing black hole candidate, Cygnus X-1, was found in 1970 using the X-ray satellite Uhuru. Cygnus X-1 is the companion of a supergiant star. The lower mass limit of objects like Cygnus X-1 can be determined from a time-resolved radial velocity curve of the companion star measured from optical absorption lines. Black hole candidates have a mass above the allowed mass range of white dwarfs (1.4 $M_\odot$, where the solar mass 1 $M_\odot$ = 1.98892 × 10^{30} kg) and neutron stars (2 − 3 $M_\odot$). The size of the black hole candidates can be inferred from X-ray variability. Objects like Cygnus X-1 belong to the class of stellar-mass black holes, which are created at the end of the life of a $\gtrsim$ 20 $M_\odot$ star via gravitational collapse. To date, a few tens of stellar mass black hole candidates in binary systems are known (Casares, 2007).

A far different class of black holes are the so-called supermassive black holes (SMBHs) with masses in the range $10^6 - 10^{10}$ $M_\odot$, which reside in the centres of galaxies. Soon after the first quasars were discovered in the 1950s and 1960s, it became clear that they are very distant objects, but their extreme brightness could not be explained by conventional energetic processes. Due to their strong and rapid variability, the light must be emitted from within a very small region. The active galactic nuclei (AGN) paradigm in which quasars are powered by material accreting onto a supermassive black hole was suggested in the 1960s by Zel'dovich (1964) and Salpeter (1964) and was further established in the 1970s (e.g., Lynden-Bell 1978). Quasars are mainly found at high

CHAPTER 1. BLACK HOLES IN GALAXIES

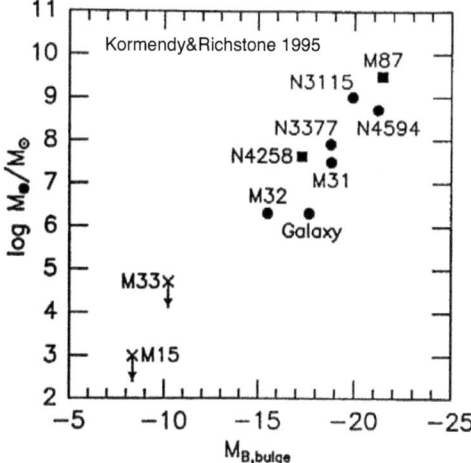

Figure 1.1: Relation between $M_\bullet$ and B-band magnitude of the bulge $M_{B,\text{bulge}}$ from Kormendy & Richstone (1995).

redshift (peak quasar activity at $z \approx 2.5$, e.g. Richards et al. 2006), thus quasar activity likely seemed to be just a phase in galaxy evolution. Hence dormant supermassive black holes ought to be found in the centres of nearby, inactive galaxies (Haehnelt & Rees, 1993; Soltan, 1982). In the 1980s, the search for these dormant black holes began. In the 1990s, the Hubble Space Telescope (*HST*) with its unprecedented high spatial resolution was launched. With the installation of the spectrograph STIS in 1997, the Golden Age started for black hole astronomers. Since then, the number of detected SMBHs increased rapidly. Nowadays it is believed that almost all galaxies harbour a SMBH at their centre.

The mass range of intermediate-mass black holes (IMBHs, $M_{\text{IMBH}} = 10^2 - 10^5$ M$_\odot$) is still largely unexplored. The existence of such a population of black holes has not yet been proven unambiguously, but there are plausible theoretical formation mechanisms (see van der Marel 2004 for a review). IMBHs could be remnants of the first generation of (presumably very massive) stars made from metal-free material (so-called Population III stars; e.g. Bromm & Larson 2004; Madau & Rees 2001), or they could be formed in dense star clusters (e.g. Gürkan et al. 2004; Rasio et al. 2004; van der Marel 2004). They possibly are black holes in the earliest stages of evolution, before they started growing. There is some evidence that some globular clusters host such an IMBH (e.g. Gebhardt et al. 2005; Noyola et al. 2008) and also some ultra-luminous X-ray sources are associated with IMBHs (e.g. Fabbiano 2004).

1.2. THE CO-EVOLUTION OF BLACK HOLES AND THEIR HOSTS

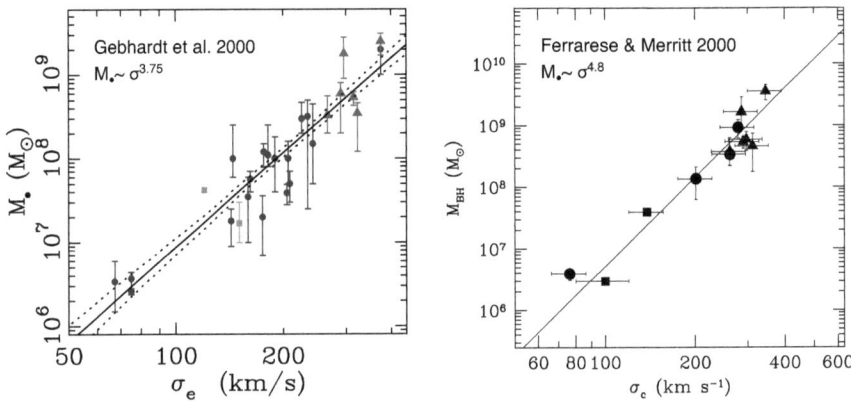

Figure 1.2: $M_\bullet$-σ relation discovered simultaneously by Gebhardt et al. (2000b) (left) and Ferrarese & Merritt (2000) (right).

1.2 The co-evolution of black holes and their hosts

With the increasing number of SMBH detections in elliptical galaxies and spiral bulges since the 1990s, it became widely accepted that all galaxies of this type harbour a SMBH in their centre (see Kormendy & Gebhardt 2001; Kormendy & Richstone 1995 for reviews). Also black hole demographics came into reach, and correlations between the mass of the central SMBH ($M_\bullet$) and galaxy properties began to emerge. A relation between the mass of the central SMBH and the luminosity or mass of the surrounding bulge component was discovered (Dressler 1989; Kormendy & Richstone 1995; Magorrian et al. 1998; see Fig. 1.1), with about 0.2 − 0.3% of the bulge mass found in the black hole. A few years later, a correlation between the mass of the SMBH and the velocity dispersion σ of the surrounding bulge was discovered independently by two groups ($M_\bullet \propto \sigma^{3.75}$, Gebhardt et al. 2000b and $M_\bullet \propto \sigma^{4.8}$, Ferrarese & Merritt 2000; see Fig. 1.2). Several other correlations with galaxy properties have been found since then, but the $M_\bullet$-σ relation appears to be the tightest one. The fact that SMBH masses are so strongly correlated with properties of the surrounding bulge implies that the formation and evolution of black holes and bulges are strongly connected.

High mass black holes ($10^8 − 10^9$ $M_\odot$) are already observed at high redshift ($z \gtrsim 6$, i.e. when the universe was only $\lesssim 1$ Gyr old) in powerful quasars. Thus there must be a mechanism via which massive black holes are grown on very short timescales, or the initial black hole mass must be very high. Accretion obviously would be such a mechanism, as it explains the high luminosity of quasars and AGN, but the question is whether it is the predominant one. $M_\bullet$-bulge correlations allow the determination of the local black hole mass density. Soltan (1982) suggested that the quasar luminosity function integrated over redshift and luminosity is related to the accreted mass of black holes during their active phase. Thus by comparing the local black hole mass density with the black hole mass density inferred from quasar light it is possible to determine whether accretion is the main black hole feeding mechanism. Such studies (e.g. Aller & Richstone 2002; Yu & Tremaine 2002) find a generally good agreement between the local and the quasar-inferred black hole mass density, but suffer from some uncertainties. The efficiency that converts mass to energy during accretion is an unknown factor. Obscured AGN are not considered in the calculation of the quasar-inferred black hole mass density. Furthermore the calculation of the local black hole mass density presently depends on the extrapolation of the $M_\bullet$-σ correlation to very low and very high masses, where no black hole mass has been measured directly, and on the assumption that the correlations are valid for all galaxy types.

Objective of this thesis

The exact role played by black holes for the formation and evolution of galaxies and the relative importance of black hole growing mechanisms are still not very well understood. The reason for that is the large effort that has to be put in the measurement of SMBH masses. Only ~ 50 galaxies currently have a measured SMBH mass. Most of these galaxies are massive, luminous ellipticals or spiral galaxies with a large classical bulge. Other galaxy types as well as low and very high-mass galaxies have been neglected so far due to their small black hole sphere of influence, low surface brightness or strong extinction due to large amounts of dust in the centre. Therefore the slope, the scatter and the universality of the $M_\bullet$-bulge relations are still uncertain and allow a lot of freedom in theories explaining the correlations between bulge evolution and black hole growth (see below). The main objective of this thesis is to clarify whether the SMBH-bulge correlations (in particular the $M_\bullet$-σ and the $M_\bullet$-L_{bulge} relations) are valid for all galaxy types and all mass ranges, and to find out which relation is the more fundamental one. Based on observations of unprecedented quality done with the new integral-field spectrograph SINFONI (Bonnet et al., 2004; Eisenhauer et al., 2003a), black hole masses in so far underrepresented galaxy categories are derived. Of main interest are (1) low-mass black holes, (2) black holes in so-called pseudobulges, where a different growing mechanism leads to bulge growth and (3) galaxies with clear signs of a recent merger event.

1.2. THE CO-EVOLUTION OF BLACK HOLES AND THEIR HOSTS

Theoretical background

There are many theoretical studies trying to explain the correlation between black hole growth and bulge evolution. These theories start from different assumptions (e.g., the nature of the accreted matter and physical processes that lead to the $M_\bullet$-bulge relations). In the models of Burkert & Silk (2001) the black holes grow via gas accretion, which is stalled when star formation in the outer accretion disc starts to exhaust the gas reservoir. They obtain $M_\bullet \propto \sigma^3$. Silk & Rees (1998) predicted the $M_\bullet$-σ relation already before it was discovered observationally. They assume $\sim 10^6$ M$_\odot$ seed black holes that accrete gas and produce an outflowing wind. This outflow interacts with the surrounding gas and thus regulates the accretion flow. This would lead to $M_\bullet \propto \sigma^5$. After strong outflows with velocities close to the escape velocity were discovered in bright quasars (Pounds et al., 2003), King & Pounds (2003) conclude that these outflows are characteristic for super-Eddington accretion and have an outflow momentum flux $\dot{M}_{out} v \simeq L_{Edd}/c$ and an outflow energy flux $0.5 \dot{M}_{out} v^2 \simeq L_{Edd}^2/(2\dot{M}_{out} c^2)$, with the mass outflow rate $\dot{M}_{out}$ and the Eddington luminosity $L_{Edd} \propto M_\bullet$. Such an outflow can be treated as a quasi-spherical wind bubble, that sweeps up gas into a shell (King, 2003). The bubble is momentum-driven as long as the shocked wind gas can be efficiently cooled, i.e., as long as the cooling time t is smaller than the Compton cooling time $t_C \simeq 10^5 R_{kpc}^2 (c/v)^2 b M_8^{-1}$ yr, where R_{kpc} is the radius of the shell in kpc, M_8 is the SMBH mass in 10^8 M$_\odot$ and b is a geometric factor. At larger radii the shell becomes energy-driven and accelerates. The velocity of the momentum-driven and the energy-driven shell depends on the Eddington luminosity L_{Edd} and thus on the mass of the black hole. At some point, when $M_\bullet$ becomes high enough, the energy-driven shell and finally the momentum-driven shell reach the escape velocity $v_{esc} \equiv \sigma$ of the galaxy and the gas accretion is stopped. King (2003) show that integrating the equation of motion of the shell inside the cooling radius twice and setting the momentum-driven shell velocity $v_m = \sigma$ leads to a relation between the SMBH mass and velocity dispersion:

$$M_\bullet = \frac{f_g \chi}{2\pi G} \sigma^4 \simeq 1.5 \times 10^8 \sigma_{200}^4 \; \mathrm{M_\odot}, \qquad (1.1)$$

with the gas fraction $f_g = \Omega_{baryon}/\Omega_{matter}$ and the electron scattering opacity χ. This relation is remarkably close to the observed relation $M_\bullet \propto \sigma^{4.02}$ (Tremaine et al., 2002). Assuming that most of the swept-up gas mass is turned into bulge stars, King (2003) derives a relation between black hole and bulge mass: $M_{bulge} \propto M_\bullet^{1.25}$. Numerous other models based on self-regulated growth have been developed up to now with different assumptions (e.g., Di Matteo et al. 2005; Marulli et al. 2008; Sijacki et al. 2007; Younger et al. 2008).

The details of the physical mechanisms that drive the self-regulation can only be constrained from observations: the more fundamental the relation between black hole mass and bulge property is, the better for the theorist. Which bulge property is the most fundamental $M_\bullet$ predictor (e.g., σ or L_{bulge}) is not yet entirely clear due to the still relative small number of direct SMBH mass measurements.

The remaining sections in this chapter give an overview about direct and indirect SMBH mass measurement methods and analyse the sample of the presently $\sim$ 50 directly measured SMBH masses. All underrepresented galaxy categories are discussed in detail concerning their location with respect to the $M_\bullet$-σ and $M_\bullet$-L_{bulge} relations.

1.3 Black hole mass measurements – status today

A large number of measurements and estimates of black hole masses using a variety of methods can be now found in the literature. The techniques used to determine the mass of a SMBH can be divided into direct and indirect methods. Direct methods measure the effect a black hole has on the dynamics of its surrounding stars or gas. They require sophisticated observations and modelling of the stellar or gas motions. The diameter of the sphere of influence (SoI) of the black hole, defined as

$$d_{\mathrm{SoI}} = \frac{2GM_\bullet}{\sigma^2}, \qquad (1.2)$$

needs to be resolved, which strongly limits the distance out to which the black hole mass of a galaxy can be measured. Indirect mass indicators are easy to measure observables which have been found to scale with $M_\bullet$ from the direct measurements. The rest of the chapter introduces the most important direct and indirect methods to measure or estimate $M_\bullet$.

1.3.1 Measurement techniques

Stellar proper motions: the Galactic centre

The Milky Way is so far the only galaxy, in which the position of the stars in the Galactic centre and their Keplerian orbits around the central mass concentration (which is likely to be identical to the radio source Sgr A*) have been measured with a very high precision (e.g. Ghez et al., 2005; Gillessen et al., 2009; Schödel et al., 2002). The most recent measurement yields a SMBH mass of $4.31 \times 10^6\,M_\odot$. Due to the high mass density present within the orbit of the star closest to Sgr A*, viable alternatives to a black hole can essentially be excluded. It is currently the best case of a SMBH detection.

1.3. BLACK HOLE MASS MEASUREMENTS – STATUS TODAY

H_2O megamasers

The emission of water maser clouds at 22 GHz from within the central parsec around a SMBH can also be used to determine the mass of a central black hole. In Seyfert 2 galaxies the dust torus around the AGN is seen edge-on, such that the large path length along the line of sight permits maser amplification. In the Seyfert 2 galaxy NGC 4258 the water maser emission describes a nearly perfectly Keplerian thin disc. The black hole mass inferred from the dynamics of the maser clouds is 3.8×10^7 $M_\odot$ (Miyoshi et al., 1995) and constitutes the second best black hole case, as due to the Keplerian motion and the small distance of the maser clouds to the nucleus the measurement errors are extremely small and essentially all other alternatives to a black hole can be excluded. Water maser emission has been detected in a number of other Seyfert 2 galaxies as well, but so far there is no other case where the rotation is perfectly Keplerian.

Stellar dynamics

In other galaxies the central region is not nearly as well resolved as in the case of the Milky Way. Instead of tracing individual stellar orbits, the integrated light of many stars can be used to reconstruct the orbital structure of the galaxy. When the spatial resolution is high enough such that the SoI of the black hole is at least approximately resolved, it is possible to reconstruct the mass of the black hole. This method is rather intricate, as both high-signal-to-noise, high-resolution imaging (to determine the stellar gravitational potential) and spectroscopy (to determine the velocity distribution of the stars) and a quite large amount of computing time are needed.

A variety of dynamical models have been developed. All assume that the gravitational potential of a galaxy has a certain shape (spherical, axisymmetric or triaxial) and that the phase-space distribution function (i.e., positions and velocities) of the stars is isotropic or can be described by two or three integrals of motion. Spherical isotropic models describe globular clusters quite well, but oversimplify real galaxies. Axisymmetric, two-integral models also do not seem to be a good representation of real galaxies, as it is assumed that the velocity dispersion is isotropic in the meridional plane (i.e., $\sigma_r = \sigma_\theta$ at every position), which is usually not observed (van der Marel et al., 1998). They deliver, however, a good first-order approximation. Axisymmetric, three-integral dynamical models are very general and a good representation of real galaxies. They are therefore widely used to measure SMBH masses and have been demonstrated to be reliable, e.g. by measuring $M_\bullet$ in the maser galaxy NGC 4258 (Siopis et al., 2009). The gravitational potential in the centre of galaxies near the black hole is usually close to spherical or, when a disc is present, axisymmetric. Deviations from axisymmetry are unlikely to play a role at small radii. As the third integral of motion is not known analytically, the orbit-based Schwarzschild (1979) method is used to build models of galaxies. A library of orbits is created in the combined gravitational potential of the stars and a black hole, and orbits are combined with weights chosen such that the kinematics and the light distribution are optimally reconstructed in the given potential. Thus, the analytic form of the third integral may remain unknown. Models for a large number of potentials (with $M_\bullet$ and the mass-to-light ratio Υ being the only difference) need to be calculated for a single galaxy. The model with the minimal χ^2 then provides the best estimate of Υ and $M_\bullet$ of the galaxy.

More general codes (triaxial, three-integral) have been developed recently (de Lorenzi et al., 2007; van den Bosch et al., 2008), but need further testing before they can be used to model black hole masses.

Axisymmetric, three-integral models are used throughout this thesis for the measurement of SMBH masses.

Gas dynamics

Many galaxies have gas in their centre, which in some cases is relaxed and confined to a rotating disc. Measuring the kinematics of this gas does not require high exposure times, as the emission lines are often very strong. The dynamics of the gas is also relatively easy to model compared to stellar dynamics, because it is assumed that the gas is rotating in a thin disc in the centre of a galaxy. The gravitational potential of the stars is determined as in the stellar dynamical case. Models of discs are then computed in this potential, again with Υ and $M_\bullet$ as free parameters, and these discs are then synthetically observed and compared with the data. The best-fitting parameters also result from a χ^2 analysis. A drawback of this method is that gas dynamics can easily be disturbed by non-gravitational forces. In addition galaxies with a regular gas distribution in the centre are quite rare.

Reverberation mapping

In the standard model of AGN (Urry & Padovani, 1995) the black hole and the accretion disc are surrounded by a region of broad emission line clouds and, further out, by a narrow emission line region. The AGN and the broad line region (BLR) are hidden by a dust torus when viewed edge-on (Seyfert 2 galaxies), but can be directly seen under lower inclination angles (Seyfert 1 galaxies) instead. The non-stellar continuum emitted by the accretion disc excites the BLR, and due to the high Doppler motions this results in broad emission lines. Changes in the continuum emission flux result in variations of the emission line flux, with a time delay that depends on the radius

1.3. BLACK HOLE MASS MEASUREMENTS – STATUS TODAY

of the BLR, $\tau = R_{\rm BLR}/c$. Using the virial theorem the black hole mass can be estimated from this time delay and the width of the emission lines, $M_\bullet = f R_{\rm BLR} \sigma^2 / G$, where f is a factor that depends on the geometry and kinematics of the BLR, which is mostly unknown. Therefore the reverberation masses are normalised to match the $M_\bullet$-σ relation (Onken et al., 2004). Although this method obviously cannot be used to determine the slope of the $M_\bullet$-σ relation, it is a very useful and reliable technique to determine $M_\bullet$ of distant type 1 AGN, where the sphere of influence cannot be resolved. It is observationally very expensive, as the AGN have to be monitored over a certain time period and a high S/N is needed to measure small flux differences.

X-ray observations

X-rays can in principle also be used to measure $M_\bullet$ in Seyfert 1 galaxies and in quiescent massive ellipticals. Broad Fe Kα line emission is present in many Seyfert 1 galaxies. Though expected to be narrow, the shape of the line is very broad and skewed. It is therefore assumed that the broad component originates in a thin and rapidly rotating accretion disc, such that Newtonian Doppler shifting causes the line being split in two, relativistic beaming enhances the blue over the red component, and gravitational redshift smears the emission into an extended red wing. The shape of the line depends on the inclination of the disc and the spin of the black hole. Using a technique similar to reverberation mapping, $M_\bullet$ can in principle be measured with the next generation of X-ray satellites. There are, however, still uncertainties in the interpretation of measured Fe Kα line shapes.

In giant elliptical galaxies the gravitational influence of the (quiescent) supermassive black hole on X-ray emitting gas in hydrostatic equilibrium results in a temperature peak. This temperature peak can be used to determine $M_\bullet$, as shown by Humphrey et al. (2008) for NGC 4649. This method is quite promising as it yields for NGC 4649 the same results as stellar kinematics, but it is probably restricted to core galaxies with gas in hydrostatic equilibrium and only the next generation of X-ray satellites can provide the necessary spatial resolution.

Application: scaling laws as indirect mass indicators

The reliability of the above methods has been tested in a few cases by measuring $M_\bullet$ with more than one technique (see e.g. Table 1.1 and Davies et al. 2006; Hicks & Malkan 2008). So far there is no reason to believe that any of the methods is systematically in error, however, the number of galaxies that allow measurements with more than one method are very limited.

Unfortunately the above mentioned direct methods can only be applied to a limited number of galaxies (only the most nearby galaxies for dynamical methods or Seyfert 1 galaxies for reverberation mapping). In addition they require large amounts of observing and computing time, thus the number of measured $M_\bullet$ only grows very slowly with time. This impedes e.g. the study of black hole evolution with redshift and generally the study of galaxy parameters with black hole mass in most galaxy samples. A way out is to use scaling relations, i.e. empirical relations found between $M_\bullet$ and other host galaxy properties, based on the sample of galaxies with direct $M_\bullet$ measurements.

The most general and well-studied relations are the correlations between $M_\bullet$ and bulge luminosity or mass (e.g. Häring & Rix 2004; Kormendy 1993, 2001; Kormendy & Richstone 1995; Magorrian et al. 1998; Marconi & Hunt 2003) and the $M_\bullet$-σ relation (Ferrarese & Ford, 2005; Ferrarese & Merritt, 2000; Gebhardt et al., 2000b; Graham, 2008b; Gültekin et al., 2009b; Tremaine et al., 2002). Furthermore correlations with the central light concentration (Graham et al., 2001; Graham & Driver, 2007) and with the bulge gravitational binding energy (Aller & Richstone, 2007) have been found. These relations have only a small scatter ($\lesssim 0.5$ dex) and can probably be applied to most galaxies. AGN, however, have a very bright nucleus and strong emission lines, which makes it difficult to measure the velocity dispersion, the luminosity or Sérsic index of the bulge component. For these galaxies there exist a number of correlations between $M_\bullet$ and emission line widths or continuum and line luminosities. The most important ones are calibrated to reverberation measurements (see e.g. McGill et al. 2008 and references therein), utilising a correlation between the BLR radius with the continuum luminosity. The uncertainties in $M_\bullet$ derived with these relations, however, are large (a factor ~ 4).

1.3.2 Catalogue of $M_\bullet$ measurements

Table 1.1 gives a list of the most reliable SMBH mass measurements to date. Only direct measurement techniques are considered. The reverberation masses are normalised against dynamical measurements and therefore not included. Upper limits and less reliable masses are included only for some bulgeless galaxies and globular clusters whose location in the extreme low-σ part of the $M_\bullet$-σ plane is particularly interesting. For a few galaxies $M_\bullet$ was measured several times using different data sets and/or different measurement methods. In the case of Cen A, only the results of Neumayer et al. (2007) and Cappellari et al. (2009) are given, as they are based on high spatial resolution 3D spectroscopy and therefore considered the most reliable estimates, but note that there are other measurements (Häring-Neumayer et al., 2006; Krajnović et al., 2007; Marconi et al., 2001, 2006; Silge et al., 2005). When two different techniques were used to measure $M_\bullet$, both results are listed in order to show the overall good agreement (NGC 3227, NGC 4258, NGC 4486, NGC 4649,

Cen A). When there were two different measurements of similar quality obtained with the same technique, both are included (NGC 1399, NGC 3377, NGC 3379). For Figs. 1.3a-g and 1.4a-f the average of the two given $M_\bullet$ was plotted for NGC 1399, NGC 3227, NGC 4486 and Cen A, the measurements of Gebhardt et al. (2003) were used for NGC 3377, NGC 3379 and NGC 4649 due to their better spatial resolution, and in the case of NGC 4258 the maser measurement was used. σ_e, i.e. the luminosity weighted velocity dispersion within the effective radius R_e is given if available. Distances were taken from Tonry et al. (2001) where available and $M_\bullet$ was adjusted to the distance where required.

1.4 Limitations of the $M_\bullet$-σ and the $M_\bullet$-L_{bulge} relation

The $M_\bullet$-σ relation found by Ferrarese & Merritt (2000); Gebhardt et al. (2000b) was based on only 26, respectively 12 SMBH mass measurements in a relatively narrow σ range (see Fig. 1.2). It is of the form

$$\log_{10}\left(\frac{M_\bullet}{M_\odot}\right) = \alpha + \beta \log_{10}\left(\frac{\sigma}{\sigma_0}\right), \quad (1.3)$$

with a normalisation constant $\alpha \sim 8$, a slope $\beta \sim 4$ and $\sigma_0 = 200$ km s^{-1}. Since the discovery of the relation the number of secure $M_\bullet$ measurements has been approximately doubled, thus new estimates of α and the slope β of the relation have been published regularly (e.g. Ferrarese & Ford 2005; Graham 2008b; Gültekin et al. 2009b; Tremaine et al. 2002). The definition of "secure" differs somewhat from author to author, resulting in $M_\bullet$-σ relations based on slightly different samples. Also the galaxy distances, definitions of σ and estimates of the error in σ vary. Presently there are ~ 50 secure direct $M_\bullet$ measurements, which are listed in Table 1.1. In addition, there are quite a number of less secure measurements and upper limits, which are not listed here (compilations of these cases are given e.g. in Graham 2008b; Gültekin et al. 2009b). Aside from the presence or absence of confidence in a few individual $M_\bullet$ measurements, the definition of the velocity dispersion σ differs between authors. In addition, σ is not measured in a consistent way for all galaxies. Some authors use the luminosity-weighted σ within a slit aperture of length $2R_e$ (Gebhardt et al., 2000b; Tremaine et al., 2002), denoted as σ_e, while others prefer σ within $R_e/8$ ($\sigma_{e/8}$, Ferrarese & Ford 2005; Graham 2008b). Often there are no measurements out to R_e or $R_e/8$, or $\sigma_{e/8}$ is estimated using Jorgensen et al. (1995). While luminosity-weighted dispersions measured in different apertures usually do not differ by more than a few percent, they may differ significantly between different data sets and/or measurement techniques. In the Circinus galaxy for example, the stellar velocity dispersion given by Oliva et al. (1995) is 168 km s^{-1}, while Maiolino et al. (1998) and Mueller Sánchez et al. (2006) found central values around 75 km s^{-1}, all using essentially the

Table 1.1: Galaxies with reliable black hole mass measurements

Galaxy	Type	AGN	$M_\bullet$ (low, high) ($M_\odot$)	σ (km s^{-1})	D (Mpc)	PSF (″)	SoI (″)	M_K mag	m	Ref.
A1836-BCG[c]	S0	FR II	3.80×10^9 (3.27,4.23)	288[8]	155.0[s]	0.05	0.52	−26.0	g	1
A3565-BCG[c]	E	FR I	1.34×10^9 (1.15,1.55)	322[8]	50.8	0.05	0.45	−26.0	g	1
Circinus[pp]	Sb	S2	1.1×10^6 (0.9,1.3)	75[o]	2.8[k]	0.0025	0.12	...	m	2
Cygnus A	E	FR II	2.4×10^9 (1.7,3.0)	270[8]	226.2[s]	0.1	0.26	−27.2[m]	g	3
IC1459	E3	R, L2	2.6×10^9 (1.5,3.7)	340[e]	29.2[t]	1.5	1.37	−25.9[m]	s	4
			3.5×10^8 (approx.)			0.1	0.18		g	4
Milky Way[pp]	SBbc		4.31×10^6 (4.25,4.37)	103[e]	0.00833	0.07	51.37	−22.3[m]	p	5
N221=M32	E2		2.5×10^6 (2.0,3.0)	75[e]	0.81[t]	0.05	0.97	−19.3[m]	s	6
N224=M31	Sb		1.4×10^8 (1.1,2.3)	160[e]	0.76[t]	0.05	12.67	−22.8[m]	s	7
N821	E6		4.0×10^7 (3.2,6.6)	209[e]	24.1[t]	0.05	0.07	−24.8[m]	s	8
N1023	SB0-		4.4×10^7 (3.9,4.9)	205[e]	11.4[t]	0.05	0.16	−23.5[m]	s	9
N1068[p]	SB	S1.9	8.2×10^6 (7.9,8.5)	151[e]	14.8[s]		0.04	−23.5[d]	m	10
N1300[p]	SBbc		7.1×10^7 (3.7,13.9)	86[e]	20.2[s]	0.05	0.84	−22.0	g	11,12
N1399[c]	cD, E1	FR I	1.2×10^9 (0.6,1.7)	317[e]	19.9[t]	0.15	1.06	−25.2[i]	s	13
			4.8×10^8 (4.1,5.5)	337[e]		0.10	0.38		s	14
N2748[pp]	SAbc	H	4.6×10^7 (0.8,8.3)	107[e]	24.4[s]	0.05	0.29	−21.3[d]	g	11,12
N2778	E2		1.5×10^7 (0.5,2.4)	175[e]	22.9[t]	0.05	0.04	−23.0[m]	s	8
N2787[p]	SB0+	L1.9	4.1×10^7 (3.6,4.5)	140[e]	7.5[t]	0.05	0.49	−21.3[m]	g	15
N3031=M81	Sab	S1.5	7.5×10^7 (6.5,9.7)	166	3.9[s]	0.05	1.24	−24.1[m]	g	16,17
N3115	S0-		9.1×10^8 (6.3,19.4)	230[e]	9.7[t]	0.21	3.14	−24.4[m]	s	18
N3227[pp]	SBa	S1.5	1.7×10^7 (0.8,2.3)	136[e]	19.8[s]	0.085	0.082	−22.2[d]	s	19
			2.3×10^7 (1.9,3.5)			0.07	0.11		g	20
N3245[pp]	S0	T2	2.1×10^8 (1.6,2.6)	205[e]	20.9[t]	0.05	0.42	−23.3[m]	g	21
N3377	E5		1.1×10^8 (1.0,2.1)	145[e]	11.2[t]	0.05	0.83	−22.8[d]	s	8
			7.9×10^7 (2.3,12.4)			0.6	0.59		s	22
N3379[c]	E1	L2/T2:	1.4×10^8 (0.6,1.7)	206[e]	10.6[t]	0.94	0.55	−23.9[d]	s	23
			1.0×10^8 (0.6,2.0)			0.21	0.39		s	24

Continued on next page

1.4. LIMITATIONS OF THE $M_\bullet$-σ AND THE $M_\bullet$-L_{BULGE} RELATION

Table 1.1 – continued from previous page

Galaxy	Type	AGN	$M_\bullet$ (low, high) ($M_\odot$)	σ (km s^{-1})	D (Mpc)	PSF (″)	SoI (″)	M_K mag	m	Ref.
N3384^p	SB0		1.7×10^7 (1.5,1.9)	143^e	11.6^s	0.05	0.13	-22.6^m	s	8
N3585	S0		3.2×10^8 (2.6,4.6)	213^e	20.0^s	0.05	0.63	-24.8^2	s	25
N3607^c	SA0	L2	1.4×10^8 (0.9,1.8)	229^e	22.8^s	0.05	0.21	-24.8^2	s	25
N3608^c	E2	L2/S2	2.1×10^8 (1.4,3.2)	182^e	22.9^s	0.05	0.49	-23.7^d	s	8
N3998	S0	L1.9	2.2×10^8 (0.5,4.2)	305	14.1^t	0.05	0.30	…	gb	26
N4026	SA0		1.8×10^8 (1.5,2.4)	180^e	13.6^s	0.05	0.72	…	s	25
N4151	SABab	S1.5	6.3×10^7 (approx.)	97^c	19.4^s	3.0	0.61	-23.5^d	s	27
			4.4×10^7 (1.2,5.6)			0.08	0.43		gb	20
N4258^p	SABbc	S1.9	3.82×10^7 (3.81,3.83)	105^c	7.3^t	0.0041	0.84	-22.4^m	m	28,29
			3.3×10^7 (3.1,3.5)			0.05	0.73		s	29
N4261^c	E2	FRI, L2	5.4×10^8 (4.2,6.6)	315^e	31.6^s	0.1	0.31	-25.2^d	gb	30,31
N4291^c	E2		3.4×10^8 (0.9,4.3)	242^e	26.2^s	0.05	0.39	-23.7^d	s	8
N4342^{pp}	S0-		2.6×10^8 (1.8,4.2)	225^e	13.4^s	0.8	0.68	-21.1^m	s	32
N4374 = M84^c	E1	FRI, L2	1.6×10^9 (1.0,2.8)	310	18.4^t	0.05	1.60	-25.1^d	gb	33
N4459	S0+	T2:	7.0×10^7 (5.7,8.3)	186^e	16.1^t	0.05	0.22	-24.5^m	gb	15
N4473^c	E5		1.2×10^8 (0.3,1.6)	190^e	15.3^s	0.05	0.39	-23.8^d	s	8
N4486 = M87^c	E0	FRI, L2	3.7×10^9 (2.6,4.7)	375^e	17.2^s	0.08	2.71	-25.4^d	gb	34
			6.7×10^9 (5.8,7.7)			0.6	4.91		s	35
N4564	E3		6.4×10^7 (5.5,6.7)	162^e	15.8^s	0.05	0.27	-23.1^d	s	8
N4596	SB0	L2::	6.3×10^7 (3.6,9.3)	152^e	13.5^s	0.05	0.36	-22.2^m	gb	15
N4649^c	E2	R	2.3×10^9 (1.5,2.8)	385^e	17.3^s	0.05	1.59	-25.5^d	s	8
			3.72×10^9 (2.66,4.46)			2.5	2.57		x	36
N4697	E6		1.9×10^8 (1.7,2.1)	177^e	11.7^t	0.05	0.92	-24.0^d	s	8
N5077^c	E3	L1.9	7.2×10^8 (4.2,11.7)	222^e	40.1^s	0.10	0.65	-24.8^2	gb	37
N5128 = Cen A	S0	FRI	5.4×10^7 (4.2,7.4)	138^e	4.2^s	0.12	1.20	-24.5^m	gb	38
			6.6×10^7 (3.0,10.0)			0.17	1.46		s	39
N5252	S0	S2	1.0×10^9 (0.5,2.6)	190°	100.8^s	0.05	0.49	-24.5^d	gb	40
N5576^c	E3		1.7×10^8 (1.3,2.0)	183^e	25.5^t	0.05	0.35	-24.2^2	s	25

Continued on next page

14

CHAPTER 1. BLACK HOLES IN GALAXIES

Table 1.1 – continued from previous page

Galaxy	Type	AGN	$M_\bullet$ (low, high) ($M_\odot$)	σ (km s^{-1})	D (Mpc)	PSF (″)	SoI (″)	M_K mag	m	Ref.
N5845	E3		2.6×10^8 (1.1,3.1)	234^e	25.9^t	0.05	0.33	−23.0^d	s	8
N6251	E0	FR I, S2	5.8×10^8 (3.8,7.7)	290^e	101.8^s	0.09	0.12	−26.0^d	g	41,31
N7052^c	E4	FR I	3.6×10^8 (2.2,6.2)	266^e	64.6^s	0.26	0.14	−25.5^d	g	42
N7457	S0		3.8×10^6 (2.3,5.0)	67^e	13.2^t	0.05	0.11	−21.8^m	s	8
N7582^{pp}	SBab	S2	5.3×10^7 (3.5,7.8)	157°	21.4^s	0.4	0.18	−22.7^d	g	43

Black holes in bulgeless galaxies and globular clusters

Galaxy	Type	AGN	$M_\bullet$ (low, high) ($M_\odot$)	σ (km s^{-1})	D (Mpc)	PSF (″)	SoI (″)	M_K mag	m	Ref.
G1	GC		1.8×10^4 (1.3,2.3)	25	0.76^t	0.05	0.07	...	s	44
M15	GC		500 (0,3000)	12	0.01	0.05	0.39	...	s	45
M33	SAcd	H	<1500	24^e	0.85	0.05	<0.005	...	s	46
N205	dE5 pec		$<2.2 \times 10^4$	21	0.74	0.115	<0.12	...	s	47
N3621^p	Sd	S2	$<3.0 \times 10^6$	43	6.6	1.0	<0.44	−15.78	s	48
N4395	SAm	S1.8	$10^4 - 10^5$	<30	4.2	...	~0.02	...	i	49
ω Cen	GC		4.00×10^4 (3.00,4.75)	20	0.0048	...	16.89	...	s	50
Pox 52	dE	S1	3.2×10^5 (2.2,4.2)	36	4.3	...	0.10	...	i	51

^{c}core galaxies (Lauer et al., 2007); ^{p}pseudobulge galaxies (Erwin, 2008; Fisher & Drory, 2008); ^{pp}possible pseudobulges (Gültekin et al., 2009b; Hu, 2008).
AGN types: S = Seyfert galaxy; L = LINER; H = HII nucleus; T = transition object between HII nucleus and LINER; R = galaxy with a strong radio core; FR I = Fanaroff-Riley type 1; FR II = Fanaroff-Riley type 2;
$^e \sigma_e$ from Tremaine et al. (2002) or the given reference; $^g \sigma_{e/8}$ from Ferrarese & Ford (2005) or the given reference; $^o \sigma$ from Maiolino et al. (1998); Mueller Sánchez et al. (2006); Nelson & Whittle (1995); Oliva et al. (1995).
^{s}Distance D from Tonry et al. (2001) or (for the Virgo galaxies only) Mei et al. (2007); ^{t}distance from the systemic velocity given by NED (Virgo+GA+Shapley) with $H_0 = 75$ km s^{-1} Mpc^{-1}; ^{k}distance from Karachentsev et al. (2007).
$^m K$-band magnitude M_K from Marconi & Hunt (2003); $^d M_K$ from Dong & De Robertis (2006) or Peng et al. (2006); $^2 M_K$ from 2MASS (Skrutskie et al., 2006)
Methods (m): p=proper motions, s=stellar dynamics, g=gas dynamics, m=maser dynamics, x=X-ray gas, i=indirect.
References: (1) Dalla Bontà et al. (2009); (2) Greenhill et al. (2003); (3) Tadhunter et al. (2003); (4) Cappellari et al. (2002); (5) Gillessen et al. (2009); (6) Verolme et al. (2002); (7) Bender et al. (2005); (8) Gebhardt et al. (2003), corrected for 9% numerical error (Gültekin et al., 2009b); (9) Bower et al. (2001); (10) Lodato & Bertin (2003); (11) Atkinson et al. (2005); (12) Batcheldor et al. (2005); (13) Houghton et al. (2006); (14) Gebhardt et al. (2007); (15) Sarzi et al. (2001); (16) Devereux et al. (2003); (17) Héraudeau & Simien (1998); (18) Emsellem et al. (1999); Kormendy et al. (1996); (19) Davies et al. (2006); (20) Hicks & Malkan (2008); (21) Barth et al. (2001); (22) Copin et al. (2004); (23) Shapiro et al. (2006); (24) Gebhardt et al. (2000a); (25) Gültekin et al. (2009a); (26) de Francesco et al. (2006); (27) Onken et al. (2007); (28) Herrnstein et al. (1999); Miyoshi et al. (1995); (29) Siopis et al. (2009); (30) Ferrarese et al. (1996); (31) Ferrarese & Ford (2005); (32) Cretton & van den Bosch (1999); (33) Bower et al. (1998); (34) Macchetto et al. (1997); (35) Gebhardt & Thomas (2009); (36) Humphrey et al. (2008); (37) de Francesco et al. (2008); (38) Neumayer et al. (2007); (39) Cappellari et al. (2009); (40) Capetti et al. (2005); (41) Ferrarese & Ford (1999); (42) van der Marel & van den Bosch (1998); (43) Wold et al. (2006); (44) Gebhardt et al. (2005); (45) van den Bosch et al. (2006); (46) Gebhardt et al. (2001); Merritt et al. (2001); (47) Valluri et al. (2005); (48) Barth et al. (2009); (49) Filippenko & Ho (2003); (50) Noyola et al. (2008); (51) Barth et al. (2004); Thornton et al. (2008).

1.4. LIMITATIONS OF THE $M_\bullet$-σ AND THE $M_\bullet$-L_{BULGE} RELATION

same near-IR spectral features (i.e., the effect of dust was the same for all measurements). While this is an extreme example, it shows that wrong σ measurements might have a large influence on the $M_\bullet$-σ relation. Circinus is one of the few galaxies at the low-mass end, thus the choice of the σ value has a large influence on the slope and on the scatter of the relationship, as shown by Gültekin et al. (2009b). When using the smaller σ, Circinus is much closer to the Tremaine et al. (2002) relationship and does not increase the scatter. Thus measurements of σ using the same technique for all galaxies would be useful.

The exact knowledge of the slope of the $M_\bullet$-σ relation is necessary to constrain the processes of bulge and SMBH co-evolution. Apart from the different slopes found due to slightly different samples or σ definitions, part of the uncertainties arise due to the fact that most σ values lie in a relatively narrow range between ~ 130 and ~ 300 km s^{-1}. Due to the small number of measurements a non-linear $M_\bullet$-σ relation would be possible, with changes of the slope at very low or very high velocity dispersions. The intrinsic scatter of the $M_\bullet$-σ relation, however, is remarkably small, only around 0.3 dex for ellipticals, but becomes larger when other galaxies are included (see Fig. 1.3d and Gültekin et al. 2009b). This poses the question whether the $M_\bullet$-σ relation is valid for all galaxies or if there are exceptions. Figs. 1.3a-g show the $M_\bullet$ vs. σ plots for the galaxies in Table 1.1 with different colour codings for different galaxy types or measurement methods in comparison to the Tremaine et al. (2002) and the Ferrarese & Ford (2005) relation.

The relation between $M_\bullet$ and bulge luminosity or mass (e.g. Erwin et al. 2004; Graham 2007; Gültekin et al. 2009b; Häring & Rix 2004; Kormendy 2001; Kormendy & Richstone 1995; Magorrian et al. 1998; Marconi & Hunt 2003; McLure & Dunlop 2002) can be written as

$$\log_{10}\left(\frac{M_\bullet}{M_\odot}\right) = a + b \log_{10}\left(\frac{L_{\text{bulge}}}{L_0}\right) \tag{1.4}$$

and has been derived for bulge luminosities in many different bands (B, V, R and K-band). Roughly 0.15% of the bulge mass is found in the SMBH. The scatter of the $M_\bullet$-L_{bulge} relation is comparable to the scatter of the $M_\bullet$-σ relation (Graham, 2007; Gültekin et al., 2009b). Difficulties in determining the slope and scatter of the relation mainly arise from the determination of the bulge luminosity. In the optical wavelength range it can be affected by dust, for spiral galaxies a proper bulge-disc decomposition is necessary, AGN emission has to be properly subtracted and last but not least it has to be clarified how to handle pseudobulges, composite bulges, bulgeless galaxies and nuclear star clusters. Table 1.1 lists the K-band magnitudes for the bulges of most galaxies (which corresponds to the total galaxy magnitude in the case of elliptical galaxies), as they are least affected by dust. Most magnitudes have been obtained from Marconi & Hunt (2003). The location of the galaxies from Table 1.1 in the $M_\bullet$ vs. L_K diagram are compared with the Marconi & Hunt (2003) relation in Figs. 1.4a-f.

1.4.1 Stars vs. gas

Different measurement techniques (stellar or gas dynamics, maser) seem to give consistent correlations with σ (Fig. 1.3a) and K-band luminosity L_K (Fig. 1.4a). This is in agreement with the SMBH masses measured with several different techniques (see Table 1.1).

1.4.2 Pseudobulges vs. classical bulges

It is now generally accepted that bulges come in two flavours: classical bulges, the scaled-down versions of elliptical galaxies, and the so-called pseudobulges, which are more similar to mini-discs than to mini-ellipticals. Kormendy (1982) was the first to notice that there are bulges that are different from the classical understanding, and later termed them "pseudobulges". Kormendy (1993) and Kormendy & Kennicutt, Jr. (2004) review properties and present a number of galaxies that host such a pseudobulge. While classical bulges in the centre of disc galaxies are thought to have formed via (minor) galaxy mergers, there is evidence that pseudobulges are a result of secular evolution. Secular evolution happens when a spiral galaxy is left alone: the gas in the disc is rearranged, i.e. it is transported from the outer regions to the inner regions via bars or spiral arms. There, it may form stars and, as it comes from the outer region of the disc, these stars slowly build up a structure in the centre of the galaxy that has properties similar to the outer disc. The identification of pseudobulges is not always easy, as their appearance can have a variety of characteristics. In addition it is possible that the two types of bulges co-exist in one galaxy (Athanassoula, 2005; Erwin, 2008; Erwin et al., 2003), thus a careful decomposition is important to identify and characterise the central bulge component. Kormendy & Kennicutt, Jr. (2004) give the currently most complete list of criteria that can be used to classify bulges as pseudobulges:

1. disc-like appearance, e.g. flattening similar to the outer disc,
2. nuclear bar,
3. box-shaped (in edge-on galaxies),
4. Sérsic index $n \approx 1-2$,
5. rotation-dominated, i.e. they fall above the line traced by an isotropic oblate rotator model (Binney, 1978) in the V_{max}/σ-ellipticity diagram,
6. low-σ outlier in the Faber-Jackson correlation (Faber & Jackson, 1976) between (pseudo)bulge luminosity and velocity dispersion,

1.4. LIMITATIONS OF THE $M_\bullet$-σ AND THE $M_\bullet$-L_{BULGE} RELATION

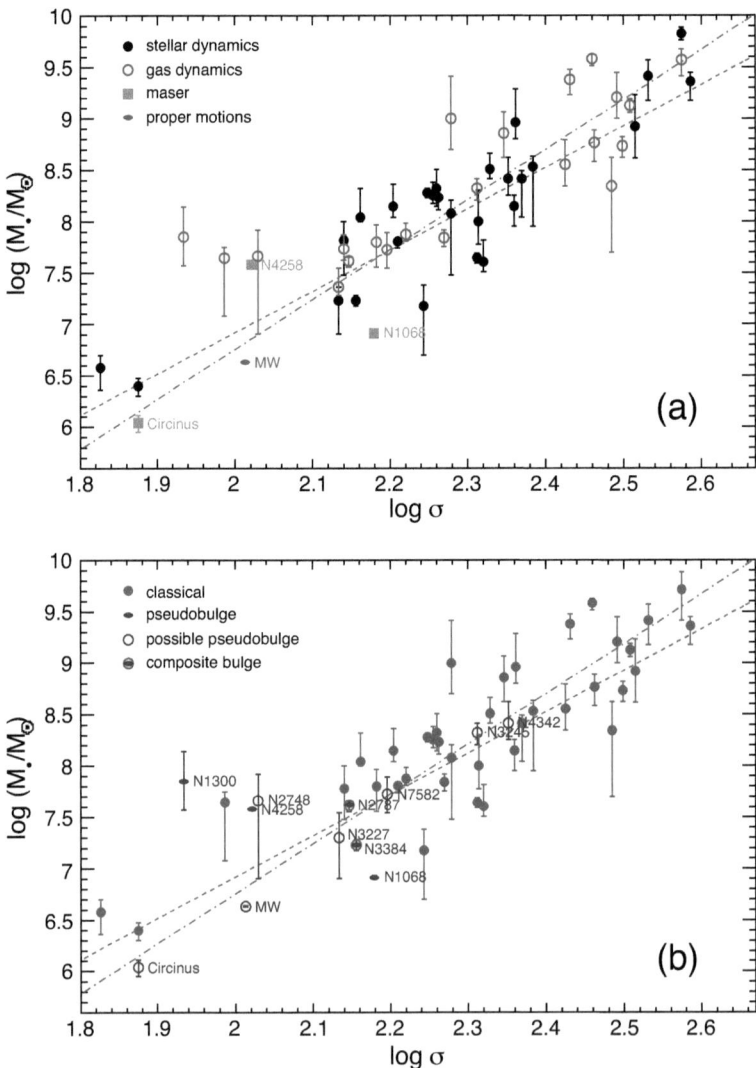

Figure 1.3: $M_\bullet$-σ relation for different subsamples. The $M_\bullet$-σ relation of Tremaine et al. (2002) (dashed line) and Ferrarese & Ford (2005) (dot-dashed line) are plotted as a reference. (a) $M_\bullet$ measured with different methods. (b) Galaxies with a classical bulge (red) and galaxies with a pseudobulge (blue).

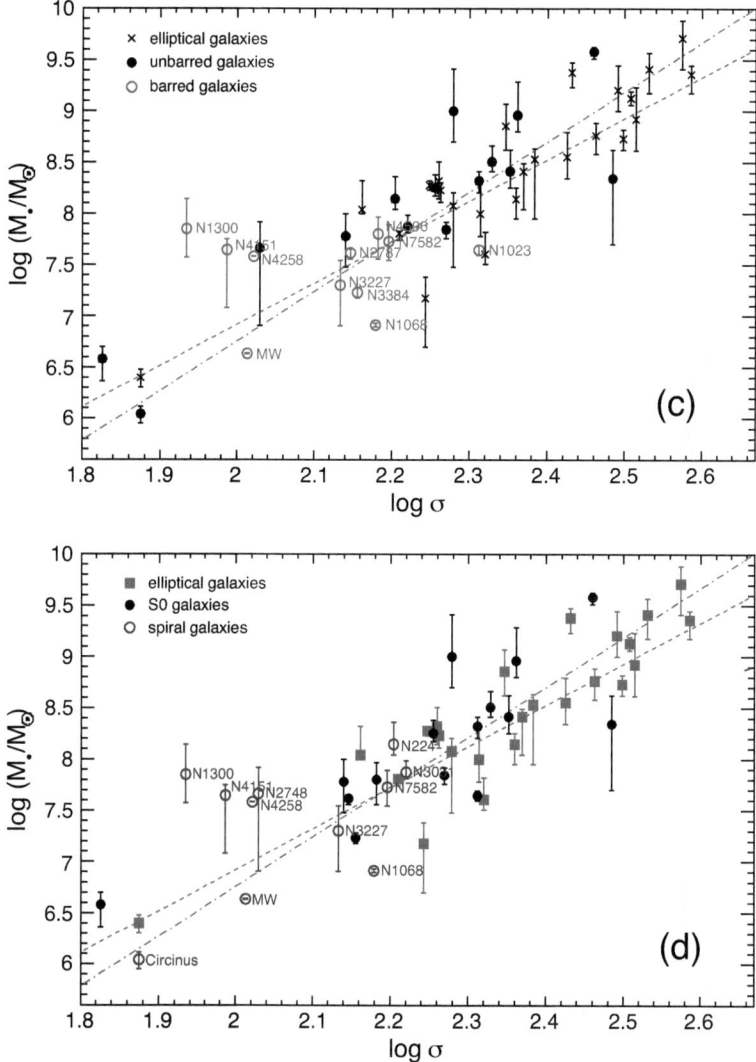

Figure 1.3: Continued. (c) Barred galaxies (red) and unbarred galaxies (black). (d) Elliptical galaxies (red), S0 galaxies (black) and spiral galaxies (blue).

1.4. LIMITATIONS OF THE $M_\bullet$-σ AND THE $M_\bullet$-L_{BULGE} RELATION

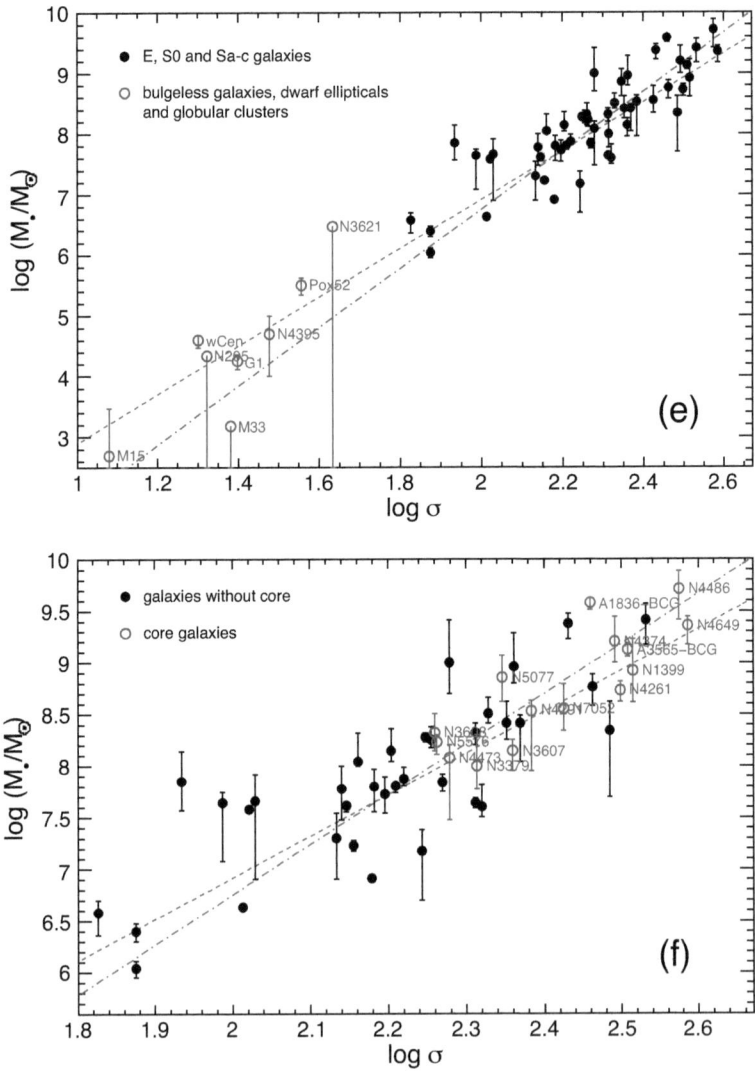

Figure 1.3: Continued. (e) Low-σ regime: bulgeless galaxies, dwarf galaxies and globular clusters. (f) Core galaxies (red) and galaxies without core (black).

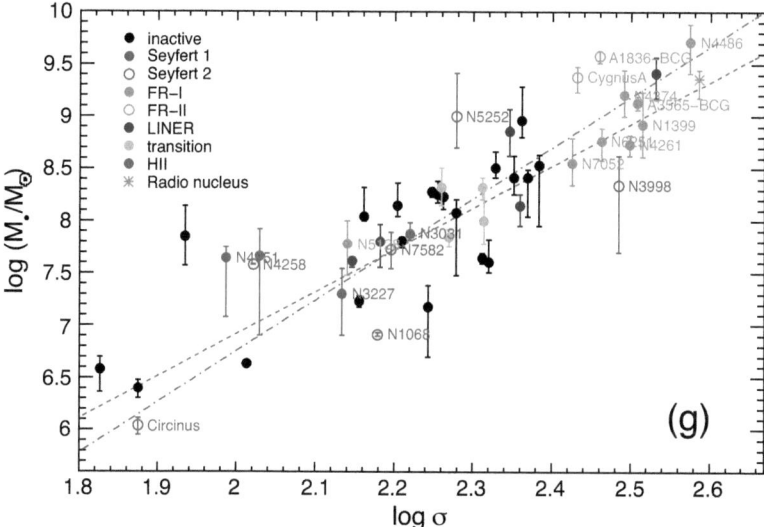

Figure 1.3: Continued. (g) Nuclear activity: inactive galaxies (black), Seyfert 1 and Seyfert 2 galaxies (red) and radio galaxies (green).

1.4. LIMITATIONS OF THE $M_\bullet$-σ AND THE $M_\bullet$-$L_{\rm BULGE}$ RELATION

7. dominated by Population I material without signs of a recent merger.

Nuclear spirals and discs (Carollo et al., 1997) or similar mid-IR colours of bulge and disc (Fisher, 2006) are also characteristics of pseudobulges. According to Kormendy & Kennicutt, Jr. (2004) a bulge is most probably a pseudobulge, if at least one of these criteria is extremely well developed. In practice, the above list is usually simplified and restricted to only a few strict criteria that need to be fulfilled by a pseudobulge. The bulge region of a galaxy, independent of the bulge type, can be identified photometrically as the excess light above the outer exponential disc. Fisher & Drory (2008) and Fisher et al. (2009) identify the bulge region as a pseudobulge, when it shows disc-like morphology (nuclear bar, nuclear spiral) and/or a Sérsic index $n < 2$. Erwin (2008) identifies pseudobulges via their flattening in comparison to the outer disc and their location above the isotropic oblate rotator line in the $V_{\rm max}/\sigma$-ellipticity diagram.

The fundamental difference between the formation mechanisms of the two bulge types provokes justified doubts about the compliance of the $M_\bullet$-σ and $M_\bullet$-$L_{\rm bulge}$ relations for pseudobulges. Are black holes present in the centres of pseudobulges at all? How is pseudobulge growth related to the growth of the central black hole? In fact, there is no reason to believe that pseudobulges are identical to classical bulges in that respect (nor that they are not). Theories of self-regulated SMBH growth (Younger et al., 2008) for example conclude that pseudobulges follow the same relation with binding energy as classical bulges do, but an $M_\bullet$-σ relation with an offset and with approximately the same slope as classical bulges. This seems to be supported by observations, as Hu (2008) finds such a normalisation offset between bulges and pseudobulges. Greene et al. (2008), on the other hand, do not observe a difference in the $M_\bullet$-σ relation between classical bulges and pseudobulges (but instead in the $M_\bullet$-$L_{\rm bulge}$ relation), and Gadotti & Kauffmann (2009) conclude that pseudobulges either agree with the $M_\bullet$-σ relation or with the $M_\bullet$-$L_{\rm bulge}$ relation, but not with both. However, the latter two studies are not based on direct $M_\bullet$ measurements. The galaxies which Hu (2008) classified as having a pseudobulge, agree only partly with the pseudobulges in Table 1.1: three out of nine pseudobulges in Hu (2008) (NGC 3079, NGC 3393 and IC 2560) do not appear in Table 1.1 due to uncertainties in the $M_\bullet$ determination (maser measurements with uncertain modelling). For these three galaxies and in addition for Circinus, the Milky Way and NGC 3227 there are some hints for the presence of a pseudobulge, but not as evident as for the other cases. NGC 2787 and NGC 3384 have been clearly identified as pseudobulge galaxies by Erwin et al. (2003) and Erwin (2004, 2008), but they also contain a classical bulge, which has an influence on the central σ and the total bulge luminosity and mass. NGC 1068 has been identified as a pseudobulge by Drory & Fisher (2007), NGC 1300 and NGC 4258 (not included by Hu 2008) have been identified as pseudobulges by Fisher & Drory (2008), but it is unclear if these are pure pseudobulges or rather composite systems. For completeness, the unconfirmed pseudobulges from Hu (2008) and Gültekin et al. (2009b) are marked as "possible pseudobulges" in Table 1.1 and Figs. 1.3b and 1.4b. Fig. 1.3b shows the location of the pseudobulges with respect to the $M_\bullet$-σ relation, Fig. 1.4b shows their location with respect to the $M_\bullet$-L_{bulge} relation. In contrast to the findings of Hu (2008), the pseudobulges are not offset from the $M_\bullet$-σ relation, but the five clear pseudobulge cases are evenly distributed above and below the relation. The scatter might be larger than for classical bulges, however. The velocity dispersions of some Hu (2008) pseudobulges (NGC 2787, NGC 3227, NGC 1068) are significantly larger than the values given in Table 1.1, which also contributes to the offset positions in the $M_\bullet$-σ plane. The pseudobulges also do not seem to deviate from the $M_\bullet$-L_{bulge} relation, but note that the total bulge magnitudes are given, which in the case of composite bulges are different from the pseudobulge magnitude.

Only five out of 50 galaxies harbour pseudobulges and additional seven may have a pseudobulge. Additionally, at least some of these pseudobulge galaxies do also host a classical component, which cannot fully be or has not been separated from the pseudobulge. Therefore it is not possible to draw conclusions regarding the locations of pseudobulges in the $M_\bullet$-σ and $M_\bullet$-L_{bulge} diagrams and the role of secular evolution regarding black hole growth. The only solution is to increase strongly the sample of pseudobulges with measured $M_\bullet$, and to clearly identify (the influence of) a possible classical component in addition to the pseudobulge in all pseudobulge galaxies. Doing this is one of the main topics of this thesis. Data for stellar dynamical modelling were taken for eleven pseudobulge galaxies and $M_\bullet$ for two of them are determined in Chapter 6.

1.4.3 Barred vs. unbarred galaxies

Graham (2008a) showed that the scatter in the $M_\bullet$-σ relation is significantly smaller when barred galaxies are excluded. The barred galaxies in his sample are systematically offset to lower $M_\bullet$. When looking at Fig. 1.3c this behaviour is not entirely reproducible. The scatter might be somewhat larger than for elliptical galaxies, but about as many barred galaxies lie above the relation as below. The samples of barred galaxies in this work and in Graham (2008a) are almost identical. NGC 4435 is not included here and NGC 1068 is not included in his sample, but that is the only difference. The offset is not seen in Fig. 1.3c because different velocity dispersion measurements for a single galaxy can show a large variety of results. The dispersions given in Graham (2008a) are partially from different sources than those given in Table 1.1. For NGC 1300 Graham (2008a) uses $\sigma = 229$ km s^{-1}, based on the value from the HyperLeda database[1] (Paturel et al., 2003). The HyperLeda value itself refers to Davies et al. (1987), where this galaxy is not mentioned, but instead this dispersion value is attributed to NGC 2300. The value of 86 km s^{-1} used in this thesis

---
[1]http://leda.univ-lyon1.fr/

instead comes from Batcheldor et al. (2005). For NGC 4151 the HyperLeda value of 156 km s^{-1} was used by Graham (2008a), but three independent measurements (see Onken et al. 2007 and references therein) give instead a value around 100 km s^{-1}. For NGC 4258 the HyperLeda value is also higher than $\sigma_e = 105$ km s^{-1} given here (from Siopis et al. 2009). For NGC 2787, the dispersion of $\sigma_e = 140$ km s^{-1} from Tremaine et al. (2002) was used in this thesis, but several other measurements in the literature quote central values around 200 km s^{-1} (see Aller & Richstone 2007), in agreement with the Graham (2008a) value. For NGC 3227 the dispersion given by Graham (2008a) is significantly larger than the dispersion from Nelson et al. (2004) used in this thesis or the HyperLeda value, and it is unclear where his value comes from. Given the very large discrepancies between σ measurements in the literature it is very hard to decide whether there are really discrepancies from the $M_\bullet$-σ relation or not for such small subsamples. A solution would be to measure σ consistently for all galaxies within the same aperture (R_e or fraction of R_e) with the same instrument (preferably an integral-field unit to avoid slit positioning errors) using the same absorption lines (preferably near-IR to avoid the influence of dust), the same fitting technique and a large template library. In addition it is necessary to measure $M_\bullet$ for more barred galaxies.

Bars are an effective tool to transport gas from the outer disc region to the centre, i.e. they are important when it comes to secular evolution (Kormendy & Kennicutt, Jr., 2004). A large fraction of the barred galaxies ($\gtrsim 50\%$) in Table 1.1 also host a pseudobulge. Therefore it is not clear if the reason for the larger scatter (and the possible offset) is caused by bars or by pseudobulges.

1.4.4 Spiral vs. elliptical galaxies

Figs. 1.3d and 1.4d show the locations of different galaxy types (ellipticals, S0 and spiral galaxies) with respect to the $M_\bullet$-σ and the $M_\bullet$-L_{bulge} relations. All galaxies seem to follow the same relations. Gültekin et al. (2009b) use a similar galaxy sample as given in Table 1.1 showing that the scatter in the $M_\bullet$-σ relation is somewhat smaller when spiral galaxies are excluded (0.31 dex compared to 0.38 dex, when Circinus is excluded from the spiral galaxies). The reason for that could be that spiral galaxies often contain bars and/or pseudobulges, and therefore might not follow the same relations with bulge properties as do ellipticals.

1.4.5 Low-σ and bulgeless galaxies

Seed black holes are assumed to form in the collapse of dense gas clouds or star clusters, or as remnants of Population III stars (Rees, 1998; Volonteri et al., 2003, 2008). Such seed black holes are difficult to detect due to their relatively small masses. They might be present in globular clusters, where intermediate-mass black holes (IMBHs) have already been discovered (e.g. Gebhardt et al. 2005; Noyola et al. 2008), or in nuclear star clusters in the centres of bulgeless galaxies. Bulgeless galaxies do not show any signs of mergers, and secular evolution processes have not yet become significant, thus they might harbour seed black holes in the earliest evolutionary stages. Some bulgeless galaxies indeed seem to host low-mass black holes, as e.g. NGC 4395 (Filippenko & Ho, 2003), which shows weak AGN emission. Others do not seem to host a black hole, as M33 (Gebhardt et al., 2001; Merritt & Ferrarese, 2001). If bulgeless galaxies do have a black hole, how does $M_\bullet$ correlate with galaxy properties? The location of the black hole masses in globular clusters and nuclear star clusters (located in bulgeless late-type spirals and dwarf galaxies) seems to be consistent with the low-mass extrapolation of the $M_\bullet$-σ relation, when using the velocity dispersion of the cluster (see Fig. 1.3e). However, at these low masses dynamical $M_\bullet$ measurements in nuclear clusters are not possible, as the SoI cannot be resolved. Thus only upper limits or indirect measurements are available so far. But not only the extreme low-mass end ($\lesssim 10^6$ $M_\odot$) is sparsely populated; there are generally very few $M_\bullet$ measurements below a few times 10^7 $M_\odot$ ($\lesssim 130$ km s^{-1}) due to the small spheres of influence which, depending on the distance, are difficult or impossible to resolve from the ground or even with *HST*. In addition, the surface brightness of these galaxies is low, such that with *HST* extremely high exposure times would be necessary. Dust also often does not allow observations of the nucleus in the optical. The low-σ range, however, is extremely important to constrain the slope of the $M_\bullet$-σ relation and to assess possible deviations from the slope between the high-mass and the IMBH mass range. Part of this thesis is therefore dedicated to the low-mass end (see Chapters 4 and 6). Using adaptive optics in the near-infrared, where the dust absorption is much less significant, helps to overcome most problems for many nearby galaxies.

1.4.6 High-σ and core galaxies

Not only the low-mass end is scarcely sampled, but also the very high-mass end of the $M_\bullet$-σ relation. There are only eight ellipticals with $\sigma > 300$ km s^{-1} with reliable $M_\bullet$ measurements, five of them with a core. According to Lauer et al. (2007) the $M_\bullet$-σ relation predicts too few extremely massive ($> 3 \times 10^9$ $M_\odot$) SMBHs compared to the predictions of the $M_\bullet$-L_{bulge} relation. Shields et al. (2006) estimate a density of relic black holes with $M > 5 \times 10^9$ $M_\odot$ of around 100 Gpc^{-1} based on the QSO luminosity function. This means that there should be an equivalent number density of galaxies with $\sigma \gtrsim 500$ km s^{-1}. No such objects have been found in the local universe. Bernardi et al. (2006) found only a few candidates with $\sigma > 500$ km s^{-1} in the Sloan Digital Sky Survey (SDSS, York et al. 2000), but most of them are probably superpositions. So where are the relics of the most massive QSOs we see at high redshifts today? Possibly the $M_\bullet$-σ relation breaks down

1.4. LIMITATIONS OF THE $M_\bullet$-σ AND THE $M_\bullet$-L_{BULGE} RELATION

at high σ. Wyithe (2006) showed that the $M_\bullet$-σ relation might be curved instead of linear in log-log space, implying that the space density of extremely massive black holes may be larger than estimated by the usually assumed log-linear form of the $M_\bullet$-σ relation. An increase of the scatter at the high-σ end could also be a possible explanation. It could also indicate, however, that the $M_\bullet$-L_{bulge} relation is the more fundamental one. It is therefore necessary to find the highest σ galaxies in the local universe and to measure their SMBH masses. The extremely high-σ objects in the sample of Bernardi et al. (2006) are unfortunately all too far away to resolve their sphere of influence. Nevertheless a breakdown or a change of slope might already be detectable with an increased number of $M_\bullet$ measurements in the accessible high-σ range.

Elliptical galaxies can be divided into two classes: core galaxies and power-law galaxies (Faber et al., 1997). Core galaxies are the result of dry mergers, where binary SMBHs fling stars away and hence produce a flattening in the central surface brightness profile, and power-law galaxies have extra light in their centres as a result of wet mergers with central starbursts (Kormendy & Bender, 2009; Kormendy et al., 2009; Merritt et al., 2007). As shown in Figs. 1.3f and 1.4e, both types of ellipticals seem to follow the $M_\bullet$-σ and the $M_\bullet$-L_{bulge} relations. According to Kormendy & Bender (2009) there exists a tight correlation between the fraction of light that is missing in core galaxies and the mass of their central SMBH, which supports the idea of the core formation mechanism through ejection of stars by the binary black hole. However, this correlation is only based on eleven Virgo ellipticals with accurate surface photometry from Kormendy et al. (2009), five of them having a reliable SMBH mass measurement. The process of core scouring leaves an imprint on the inner orbital structure (Quinlan & Hernquist, 1997). Only few SMBH masses in core galaxies have been measured using stellar kinematics, and implications on the orbital structure are ambiguous (Cappellari et al., 2008; Gebhardt et al., 2003). More stellar dynamical studies would therefore be needed. High-σ and core galaxies are not part of this thesis, but part of a larger survey with SINFONI, which is why data for several of these galaxies have been taken (see Chapter 2).

1.4.7 AGN vs. inactive galaxies

Several forms of nuclear activity are present in the galaxies of Table 1.1. The spectra of H II nuclei, low-ionisation nuclear emission-line regions (LINERs) and Seyfert galaxies can be distinguished by certain emission line ratios (Ho et al., 1997). Emission lines can be excited via photoionisation by hot, young stars (H II nuclei) or by AGN continuum radiation (Seyfert nuclei). In LINERs both stars or a low-luminosity AGN could be a possible source (Maoz, 2007). A differentiation between H II and LINER galaxies on the basis of emission-line ratios is sometimes inconclusive (transition objects). Young star clusters or on-going star formation are not restricted to H II and

CHAPTER 1. BLACK HOLES IN GALAXIES

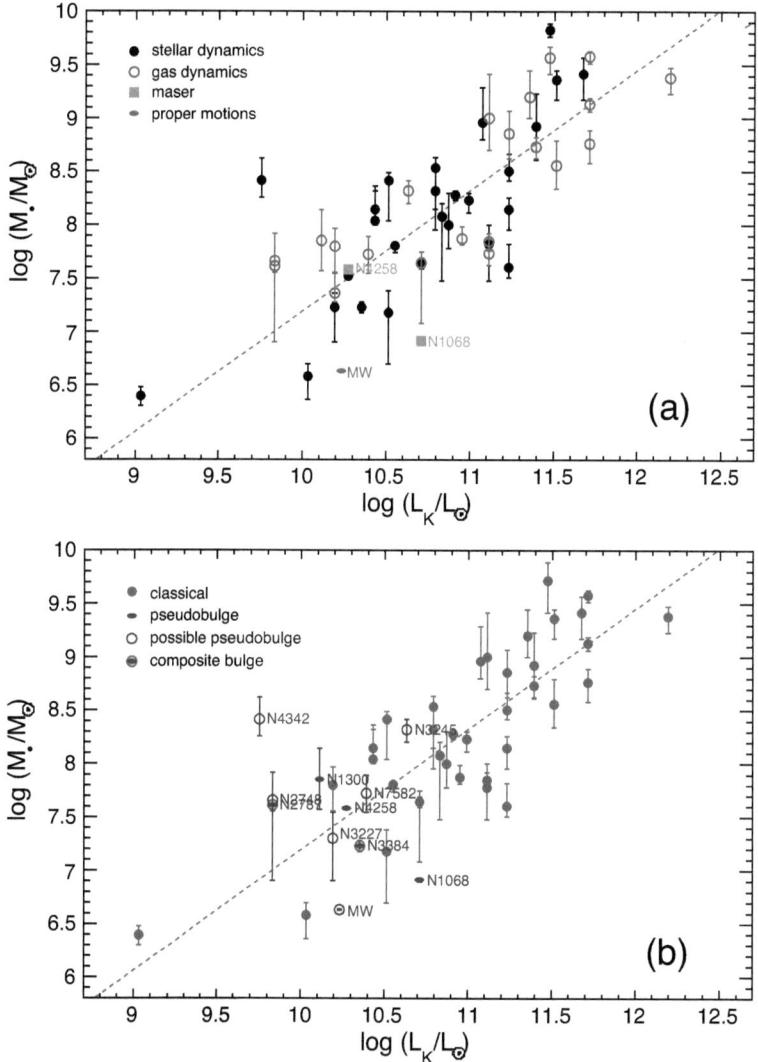

Figure 1.4: Relation between black hole mass $M_\bullet$ and K-band luminosity of the bulge L_K for different subsamples. The $M_\bullet$-L_K relation of Marconi & Hunt (2003) (dashed line) is plotted as a reference. (a) $M_\bullet$ measured with different methods. (b) Galaxies with a classical bulge (red) and galaxies with a pseudobulge (blue).

1.4. LIMITATIONS OF THE $M_\bullet$-σ AND THE $M_\bullet$-$L_{\rm BULGE}$ RELATION

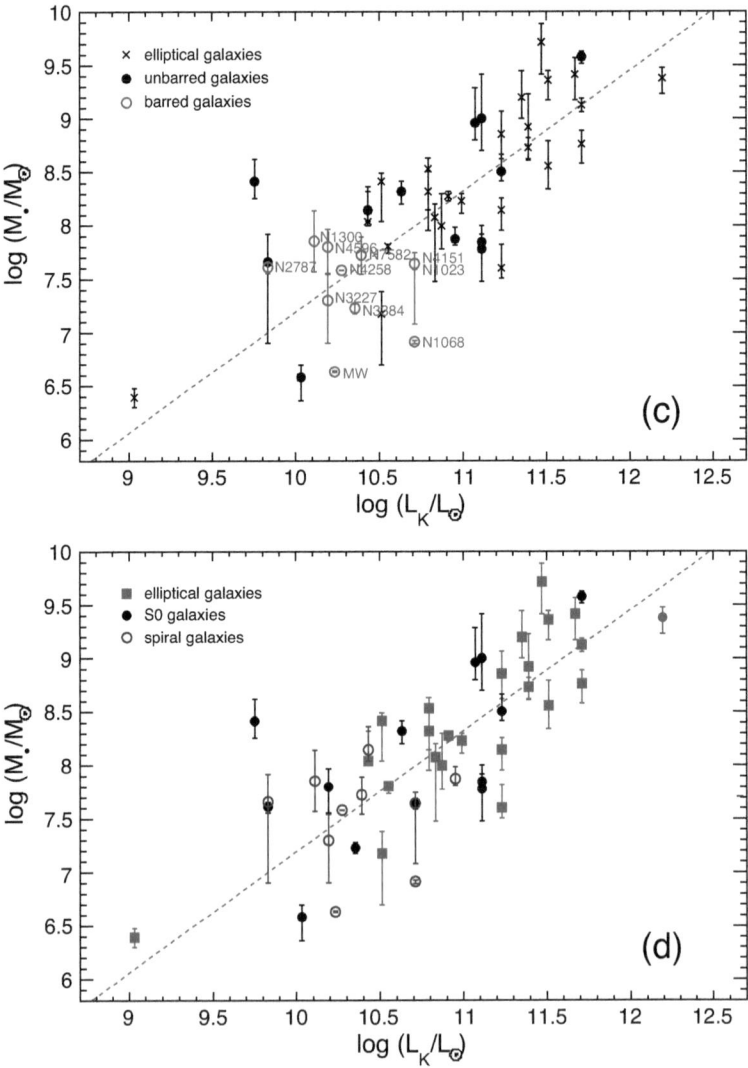

Figure 1.4: Continued. (c) Barred galaxies (red) and unbarred galaxies (black). (d) Elliptical galaxies (red), S0 galaxies (black) and spiral galaxies (blue).

CHAPTER 1. BLACK HOLES IN GALAXIES

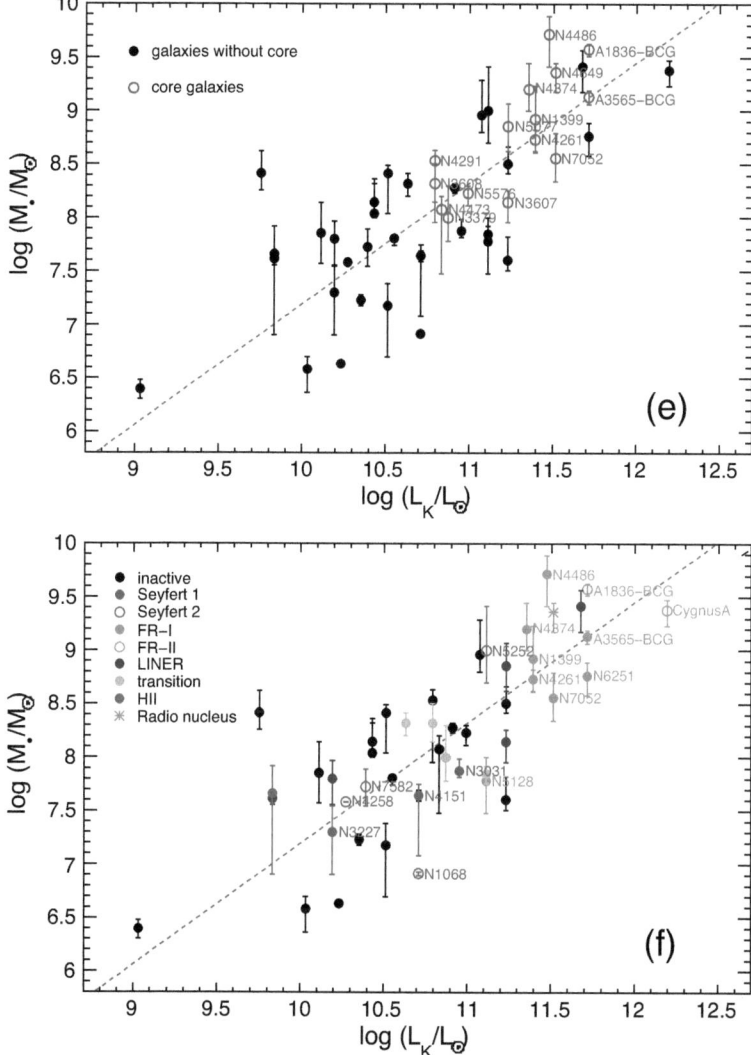

Figure 1.4: Continued. (e) Core galaxies (red) and galaxies without core (black). (f) Nuclear activity: inactive galaxies (black), Seyfert 1 and Seyfert 2 galaxies (red) and radio galaxies (green).

1.4. LIMITATIONS OF THE $M_\bullet$-σ AND THE $M_\bullet$-L_{BULGE} RELATION

LINER galaxies, but are often observed in Seyfert galaxies (e.g., Gu et al. 2006; Joguet et al. 2001). Seyfert galaxies and LINERs can be further divided into "type 1" (broad permitted emission lines) and "type 2" (narrow emission lines) galaxies according to the standard model of AGN (Urry & Padovani, 1995). Broad emission lines originate in high-velocity clouds close to the AGN (the "broad line region" or BLR), which are hidden behind a dust torus when the AGN is viewed edge-on (type 2) and which cause broad emission lines in AGN viewed under lower inclination angles (type 1). Seyfert nuclei exhibit, like QSOs, relatively high accretion rates of several percent or more of the Eddington limit. This is likely to be the main mechanism of black hole growth today in the population of low-mass black holes ($\lesssim 10^8$ $M_\odot$, Heckman 2008; Heckman et al. 2004).

Radio galaxies differ from Seyfert galaxies in many aspects. They are usually giant elliptical galaxies, which host the most massive black holes (Best et al., 2005). The accretion rates are much lower than in Seyfert galaxies. Typical for radio galaxies are highly collimated relativistic outflows (jets) and broad radio lobes around the path of the jets. The morphology of jets and lobes is used to distinguish between "FR I" (surface brightness decreases with the distance from the nucleus) and "FR II" (bright hot spots at the outer edges of the radio lobes) radio galaxies (Fanaroff & Riley, 1974). Radio galaxies are usually located at the centre of clusters or groups, where they are assumed to regulate the cooling of hot gas (Croton et al., 2006).

Most low and intermediate-σ ($\lesssim 250$ km s^{-1}) galaxies in Table 1.1 show only weak Seyfert or LINER type nuclear activity or are inactive. In the high-σ range most galaxies are FR I radio galaxies (see Fig. 1.3g and 1.4f). It is very important to determine the validity of the $M_\bullet$-σ and the $M_\bullet$-L_{bulge} relations for AGN. AGN activity is a sign of active black hole growth, therefore studying black hole and bulge properties in galaxies with AGN today and at higher redshifts might shed light on the growing mechanisms and co-evolution of black holes and the surrounding bulges, and hence explain the $M_\bullet$-σ and $M_\bullet$-L_{bulge} relations. Gas inflow due to merger events or secular evolution are the likely causes that trigger AGN activity. However, even bulgeless galaxies like NGC 4395, that show no signs of a merger event or secular evolution, can have an active nucleus. Due to the strong non-thermal and line emission of Seyfert/QSO-type AGN, they can be detected out to very high redshifts. Thus, in order to estimate $M_\bullet$ of distant AGN reliably, the validity and scatter of the $M_\bullet$-σ and other scaling relations has to be analysed, such that for high-z AGN the most reliable observable can be used to estimate $M_\bullet$.

Local Seyfert galaxies with $M_\bullet$ measured via reverberation mapping (Onken et al., 2004) and Seyfert galaxies from SDSS (Greene & Ho, 2006) seem to follow the $M_\bullet$-σ relation, though with a larger scatter and possibly a lower zero point and shallower slope than inactive galaxies (Woo et al., 2008). However, it cannot be excluded that the cause of these differences is not the AGN but the nature of the host galaxy, which in the case of Seyfert type AGN are usually spiral galaxies, possibly with bars. A few SMBH masses in Seyfert galaxies have been measured directly using stellar or gas kinematics or maser emission. Two of them (NGC 3227 and NGC 4151) also have reverberation masses. The dynamical measurements are in agreement with the reverberation measurements, and the entire sample of Seyfert galaxies in Table 1.1 shows a similar trend as in Woo et al. (2008) (see Fig. 1.3g). While more direct $M_\bullet$ measurements for Seyfert galaxies would be desirable, it is also highly important to analyse the behaviour of late-type galaxies, barred galaxies and pseudobulges in detail in order to identify the object category that is responsible for the observed larger scatter and possible offset from the scaling relations.

Mergers likely trigger AGN activity and thus black hole growth. Galaxies that show clear signatures of a recent merger event are barely studied respective their black hole masses (the only example is the radio galaxy Cen A). Depending on the timescales and simultaneity of bulge growth and black hole growth, bulge growth might lag black hole growth or vice versa. The merger history (major or minor, one or several mergers, gas-rich or gas-poor, involved galaxy types and mass ratios, etc.) might also be an important factor in the evolution of bulge and black hole during such a merger event. For Cen A many different $M_\bullet$ measurements are published (Cappellari et al., 2009; Häring-Neumayer et al., 2006; Krajnović et al., 2007; Marconi et al., 2001, 2006; Neumayer et al., 2007; Silge et al., 2005). While the first measurements partly seem to deviate significantly from the $M_\bullet$-σ relation, the most recent results based on high-resolution integral-field data do not. As mergers trigger AGN activity, and at higher redshift an increasing fraction of galaxies have morphologies indicative of recent merger events, it is particularly important to find out whether the scaling relations used for "normal" galaxies are also valid for merger remnants. In Chapter 5, $M_\bullet$ of the merger remnant and FR I radio galaxy Fornax A is measured and conclusions regarding the validity of the $M_\bullet$-σ and $M_\bullet$-L_{bulge} are drawn.

1.5 Outline of the thesis

There are still many open questions regarding the $M_\bullet$-bulge relations, in particular at the low and at the high-mass end, for pseudobulges, barred galaxies and merger remnants. The behaviour of the $M_\bullet$-bulge relations and possible differences between galaxy types can tell us about the relative importance of different growing mechanisms. Stellar dynamics can be generally used for all galaxies to reliably determine $M_\bullet$ regardless of the galaxy type, as long as the sphere of influence is resolved. In the past, however, it was impossible to obtain adequate data for the derivation of stellar dynamical SMBH masses in many galaxies except for the most nearby brightest ellipticals

1.5. OUTLINE OF THE THESIS

or early-type spirals. Low-mass black holes have a very small sphere of influence well below $1''$. In almost all low-σ, late-type and pseudobulge galaxies and AGN dust obscuration is a serious problem. To measure the stellar kinematics a S/N of at least 30 per pixel is needed, which for most instruments means unrealistically high exposure times if the galaxy is faint (i.e. almost all low-mass galaxies and core galaxies). In AGN the non-stellar emission of the active nucleus often dilutes the stellar absorption features. A large fraction of the black hole mass measurements performed so far are based on *HST* STIS spectra. They have a very high spatial resolution, but as STIS is an optical instrument and the light collecting power of *HST* is quite low, it could only be used for massive, bright galaxies without a significant amount of dust. Only since the advent of adaptive-optics assisted near-infrared instruments at large ground-based telescopes (e.g. SINFONI at the Very Large Telescope or NIFS at Gemini North) a few years ago it is now possible to overcome most of the difficulties for a reasonable number of interesting objects. It is now possible to reach a diffraction-limited spatial resolution comparable to *HST* from the ground, the large light collecting power of 8m-class telescopes reduces the needed integration time for faint objects to a fair amount, and dust obscuration plays only a minor role in the near-infrared. The instrument SINFONI (see Chapter 2) was built by the Max-Planck-Institut für extraterrestrische Physik (MPE), the European Southern Observatory (ESO) and the Netherlands Research School for Astronomy (NOVA). As our group (the Opinas group at MPE, led by Prof. Dr. Ralf Bender), was involved in the development of SINFONI, seven nights of guaranteed time observations (GTO) were awarded to us by ESO. Thanks to Ralf Bender all seven nights were assigned to the black hole project, which made this thesis possible. I got the opportunity to observe low-σ and pseudobulge galaxies and a merger remnant with SINFONI and to measure the mass of the central black holes. More recently also data for high-σ and core galaxies were taken with SINFONI by our group, but they will not be discussed in this thesis.

In Chapter 2 the instrument SINFONI is introduced and the selection of the galaxies and the observations are described. The derivation of the stellar kinematics from SINFONI spectra is explained in detail in Chapter 3. The SINFONI data, the derived kinematics, the photometry of the galaxies and the stellar dynamical modelling of the black hole masses are presented in Chapter 4 for the low-σ galaxy NGC 4486a, in Chapter 5 for the merger remnant and radio galaxy Fornax A (NGC 1316) and in Chapter 6 for the composite pseudobulges NGC 3368 and NGC 3489. Chapter 7 summarises the results, shows the implications on the $M_\bullet$-σ and $M_\bullet$-L_{bulge} relations of these four galaxies and outlines the present status of the SINFONI observations, the data analysis and the next steps.

2
Observations and data reduction

Traditionally spectroscopy is done by dispersing the light of an object that falls through a narrow slit. While this is very useful in general, it becomes rather unsatisfactory e.g. when studying the dynamical structure and stellar populations of galaxies. To obtain a full set of information about a galaxy (i.e. photometry and kinematics) is impossible with pure longslit spectroscopy. Based on longslit kinematics only, multiple kinematic components, central bars, discs or counter-rotating cores are very difficult to detect. Placing the slit on many positions of the galaxy (e.g. along the major and the minor axis and some angles in between) would be a possible solution, but this is extremely expensive in terms of exposure time and does not allow to obtain spatially contiguous spectra. The positioning of a slit is always connected with a small error, which can result in differences between the kinematics or black hole mass measurements that are based on different longslit data. It therefore can be dangerous to determine black hole masses based on only one longslit position. Also the orbital distribution of a galaxy can only be unambiguously determined if the velocity distributions at all spatial positions are known.

During the last decade, the development of a novel type of spectrographs, the so-called integral-field spectrographs (IFS), progressed a lot and nowadays almost all large telescopes are equipped with such instruments. IFSs overcome all major disadvantages of longslit spectroscopy. They are able to obtain spectra at all spatial resolution elements ("spaxels") in a 2-dimensional field in a single shot and the result is a 3-dimensional datacube (the spatial coordinates x and y and the wavelength λ). Compared to other 3D techniques like Fabry-Perot interferometry IFSs have the advantages that the spectral range is much larger and that all data are obtained simultaneously.

2.1 SINFONI

The instrument used for all observations in this thesis is SINFONI (Spectrograph for INtegral Field Observations in the Near Infrared, Bonnet et al. 2004; Eisenhauer et al. 2003a), which consists of the integral-field unit SPIFFI (SPectrometer for Infrared Faint Field Imaging, Eisenhauer et al. 2003b) and the adaptive optics module MACAO (Multi-Application Curvature Adaptive Optics, Bonnet et al. 2003) and is mounted at the fourth unit telescope (UT4) of the Very Large Telescope (VLT) in Chile. SPIFFI (see Fig. 2.1) belongs to the class of image slicers. Fig. 2.2 shows the basic working principle of the SPIFFI image slicer. The field of view, which can be chosen between $0.8'' \times 0.8''$, $3'' \times 3''$ and $8'' \times 8''$, is slit into 32 slitlets of 64 resolution elements each by a set of 32 mirrors. These slitlets are then combined by a second set of 32 mirrors to a pseudolongslit with a width of 25 mas, 100 mas or 250 mas (1 mas $\equiv 10^{-3}$ arcseconds), depending on the field of view chosen, and then dispersed by a grating. There are four gratings available, a J-band (1.1 − 1.4 µm), an H-band (1.45 − 1.85 µm), a K-band (1.95 − 2.45 µm) and a combined $H + K$-band (1.45 − 2.45 µm) grating. The dispersed light is then directed to a Rockwell 2k×2k pixel Hawaii 2RG detector. The spectral resolution $R = \lambda/\Delta\lambda$ is around 2000, 3000, 4000 and 1500 in J, H, K and $H + K$ respectively. With the data reduction software (Abuter et al., 2006; Modigliani, 2009; Schreiber et al., 2004) a three-dimensional datacube is reconstructed from the two-dimensional spatial information and the spectral information (32 slitlets of 64 spaxels length ×2048 pixels in wavelength direction). This datacube can be used to analyse the spectra at each position in the field of view, and to analyse the photometry by collapsing the cube along the wavelength direction. SINFONI has been commissioned in 2004 and is officially operational since April 1st, 2005. In the following the smallest selectable field of view ($0.8'' \times 0.8''$ with spaxels of the size $0.0125'' \times 0.025''$) will be referred to as the "25mas scale", the intermediate field of view ($3'' \times 3''$ with spaxels of the size $0.05'' \times 0.1''$) as the "100mas scale" and the largest field of view ($8'' \times 8''$ with spaxels of the size $0.125'' \times 0.25''$) as the "250mas scale".

2.1.1 Adaptive optics

The VLT has a diffraction-limited resolution of $\sim 0.06''$ in the K-band, but this resolution could only be reached if the VLT was, like *HST*, located in space. Here on Earth the atmosphere strongly

CHAPTER 2. OBSERVATIONS AND DATA REDUCTION

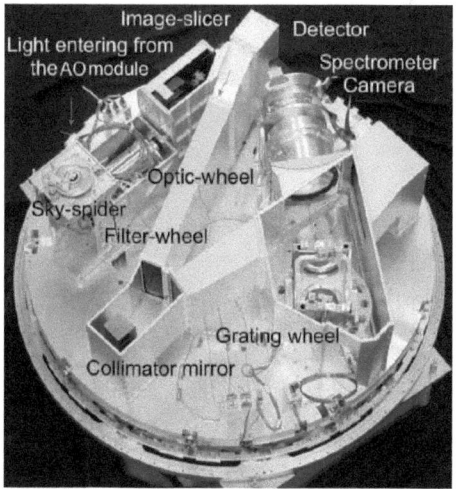

Figure 2.1: Inside view of SPIFFI (taken from Dumas 2008).

reduces the spatial resolution due to turbulences caused by temperature fluctuations, which induce changes of the air refractive index. Therefore astronomical images taken at ground-based telescopes are always blurred, i.e. their spatial resolution is reduced. The seeing (i.e. the strength of the blurring) and hence the spatial resolution is measured by the FWHM of a point source image. Depending on the weather conditions it ranges between a few tenths and a few arcseconds. The effects of the atmosphere can be corrected by the powerful technique of adaptive optics (AO). Part of the AO module MACAO is a dichroic beam splitter, which reflects the light from a reference star (usually a bright star very close to the object of interest, or the object itself) in the optical wavelength range towards an optical curvature wavefront sensor (WFS), while the near-infrared light is transmitted to the spectrograph SPIFFI. The WFS then measures the distortions of the optical wavefront from that reference object. This information is used by a real-time computer, which controls a deformable mirror, to calculate its optimal shape. The mirror is then shaped appropriately and corrects the shape of the wavefront of the object of interest.

The performance of the AO system strongly depends on the atmospheric conditions, in particular on the atmospheric coherence length r_0 and the coherence time τ_0. r_0 is the length over which perturbations of the wavefront are correlated. It decreases as turbulence increases and additionally it depends on the wavelength as $r_0 \propto \lambda^{6/5}$. In the optical r_0 is around $10-20$ cm, thus in the K-band it is about 6 times larger. This is the reason why AO corrections are easier to do in the near-IR

2.1. SINFONI

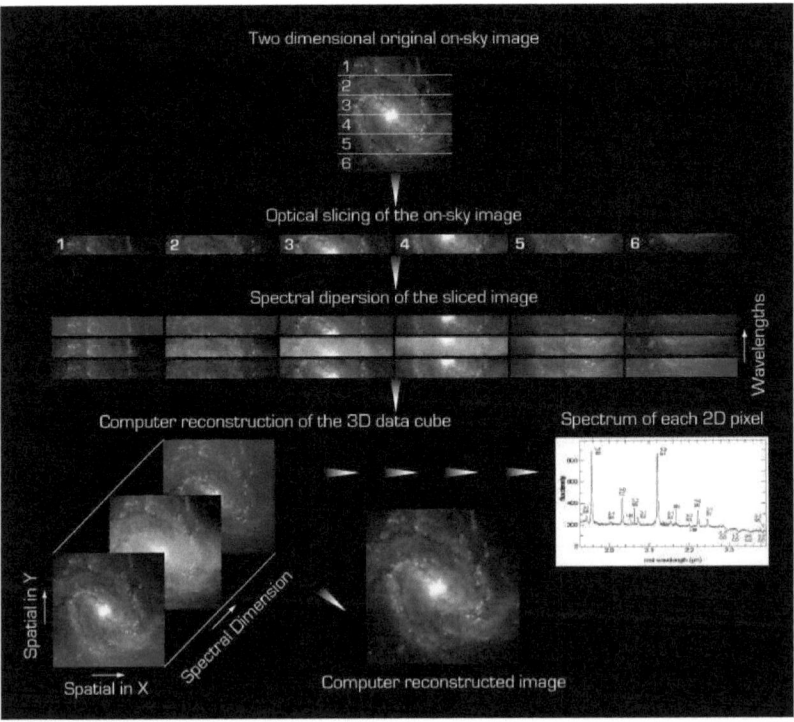

Figure 2.2: Principle of integral-field spectroscopy with an image slicer type instrument (taken from Dumas 2008)

CHAPTER 2. OBSERVATIONS AND DATA REDUCTION

than in the optical, and why the seeing, which is approximately λ/r_0 (for $r_0 < D$, the diameter of the primary mirror of the telescope), decreases with increasing wavelength. The coherence time is defined as $\tau_0 = 0.31 r_0/v_0$, where v_0 is the average wind speed in the atmosphere. In the optical τ_0 is usually of the order of a few milliseconds. The larger r_0 and thus τ_0, the better will be the AO performance. The main parameter that characterises the quality of the AO corrected image is the Strehl ratio, i.e. the ratio of the peak intensity of an observed image to the peak intensity of a perfect image at the diffraction limit with the same flux. With SINFONI the highest Strehl ratios ($\lesssim 0.6$ for perfect atmospheric conditions using an $R = 9$ mag bright reference object at $0''$ distance from the object of interest) can be reached in K-band. The AO correction in the H-band and the J-band is significantly worse.

The reference source used to measure the wavefront distortions can be a real object (referred to as "natural guide star" or NGS), like a nearby star, or an artificial star, created by a laser ("laser guide star" or LGS). Ideally an NGS should be a star with a magnitude brighter than $R \sim 14$ mag, but nuclei of galaxies (AGN are usually bright and pointlike) can be used as well, if $R \lesssim 14$ mag within a $3''$ diameter aperture. In addition an NGS can only be chosen within the size of the isoplanatic patch, which is defined in terms of an isoplanatic angle $\theta_{iso} = r_0/H$, where H is the distance to the turbulent layer in the atmosphere. In the optical θ_{iso} is only a few arcseconds in diameter, while in K-band it is around $20''$. Thus when observing with SINFONI in the K-band, the NGS should be chosen in a region closer than $10 - 15''$ to the object of interest. Beyond that, the performance of the AO correction degrades strongly. Partial correction can still be obtained for NGS magnitudes $14 < R < 17$ mag and distances out to $30''$.

2.1.2 Laser guide star

Adaptive optics with NGS can only be used on a tiny part of the sky ($\lesssim 1\%$), namely only the brightest objects and areas closer than a few arcseconds to them. Late-type galaxies and pseudobulges are usually faint, dusty and often do not have an active nucleus. Galactic nuclei with a bright star located within a few arcseconds of the nucleus are very rare (only a handful of these objects are known in the entire sky). Therefore a system has been developed that creates an artificial star by pointing a laser to a spot in a region of the atmosphere high above the telescope. There are two types of these laser guide stars (LGS): sodium and Rayleigh beacon guide stars. Rayleigh beacons focus the laser at low-altitude ($10 - 20$ km) regions of the atmosphere and measure the Rayleigh backscatter. Sodium beacons focus a sodium laser (Na I D resonance lines at 589.2 nm) at the thin layer of atomic sodium at ~ 90 km altitude in the mesosphere. The atomic sodium is excited and then re-emits fluorescence light at the same wavelength. In contrast to Rayleigh beacons they are

2.1. SINFONI

relatively complicated and expensive to construct, but provide the better AO correction as a higher region in the atmosphere is used. The LGS installed at the VLT UT4, PARSEC (Bonaccini et al., 2002; Rabien et al., 2004), is a sodium type laser designed to produce an artificial star of $V \approx 11$ mag for AO correction. It has been commissioned in early 2007, but was only partly operational for a while due to maintenance problems. A final commissioning has been done in mid 2008. Since then it is fully operational, but as a "laser specialist" is needed to ensure the usability, it is not always in operation and usually only offered in service mode and for observers with guaranteed time observations (GTO).

The apparent position of the observed object changes with time due to atmospheric turbulence. The laser light passes the atmosphere twice and is therefore randomly perturbed on the upgoing *and* on the downgoing path. Thus the position of the laser is not a good reference for the variations of the arrival angle (tip-tilt) of the object light produced by the atmosphere. Therefore the LGS is used in combination with a natural guide star (so-called tip-tilt star or TTS) to measure the tip-tilt component. The TTS can be fainter ($R \leqslant 18$ mag) and further away ($< 1'$) than the reference star used for NGS observations (though as in NGS mode the AO correction degrades with decreasing brightness and increasing distance of the TTS). Thus the sky cannot be fully covered, but the coverage is much larger than for the NGS mode (50% compared to 1%). It is also possible to use PARSEC without a tip-tilt star and thus to increase the sky coverage up to 100%, but the AO correction will only be modest (though clearly better than for no-AO observations).

In LGS mode it is not possible to reach Strehl ratios as high as in NGS mode. This has several reasons. First of all the artificial star is generated at a finite altitude and is not located at infinity as astronomical sources are. Thus the light from the artificial star forms a conical beam, while the atmospheric volume in which the wavefront of the astronomical object is distorted is cylindrical (this is called the "cone effect" or "focus anisoplanatism"). Thus less turbulence than that actually affecting the wavefront is measured, which reduces the achieved Strehl ratio. Another problem is that the image of the artificial star is not pointlike due to the width of the sodium layer of ~ 10 km. Other factors that may decrease the image quality are the variable properties of the sodium layer on one side and the fine tuning of the laser to compensate for that on the other side. The laser power has to be high enough to excite a sufficient fraction of Na atoms, but it should not be too high in order to avoid saturation. The fraction of excited Na atoms at a given laser power can be maximised by modulating the flux of the laser such that the laser line width matches the width of the atmospheric Na, which in turn depends on the thermal motions of the Na atoms. Thus if the Na layer properties vary significantly during the observations, or the laser power and line width cannot be adjusted adequately, the returned power is low and the image quality is degraded. In addition it is not possible to use the LGS when there are cirrus or other clouds.

2.2 Selection of galaxies

The selection of galaxies to be observed depends on several aspects:

- object category,
- availability of high-resolution imaging (e.g. *HST*) and kinematic information,
- distance and estimated sphere of influence,
- inclination,
- K-band surface brightness,
- observability,
- availability of an NGS or TTS,
- availability of the LGS.

All selected galaxies that have been observed so far are listed in Table 2.1 together with some of their properties. Most of them belong to one of the categories discussed in the last chapter. Pseudobulges or composite bulges and low-σ galaxies with only a small classical bulge are of main interest in this thesis. Thus it is important that high-resolution imaging of the pseudobulge region (e.g. from *HST*, preferably in several bands) and some kinematic information (v and σ along the major axis) are available which can verify the presence of a pseudobulge or composite bulge in a certain galaxy. Samples of pseudobulges or disc galaxies with candidate pseudobulges used for the selection of galaxies were taken e.g. from Carollo et al. (1998), Kormendy & Kennicutt, Jr. (2004), Erwin (2004), Drory & Fisher (2007), Fisher & Drory (2008) and Fisher et al. (2009). For all other galaxies (in particular low-σ and late-types) it is also important to have some photometric and kinematic information in order to (1) have an idea of the velocity dispersion and thus the angular size of the sphere of influence, and (2) distinguish them from pseudobulges. Low-σ galaxies can be found in collections of catalogues of galaxies provided by VizieR,[1] a list of bulgeless galaxies is given e.g. by Böker et al. (2002). Accurate photometry is also required to determine the stellar mass density for the dynamical modelling, and kinematics outside the SINFONI field of view may help to constrain the SMBH mass.

Each selected galaxy has a well-known distance (e.g. from Tonry et al. 2001), such that together with the velocity dispersion of the galaxy the mass of the central black hole (via the $M_\bullet$-σ relation of Tremaine et al. 2002) and thus the diameter of the sphere of influence can be estimated. Dynamical modelling also requires to know the distance and thus the angular scale of the object as accurate

---

[1]VizieR catalogue access tool, CDS, Strasbourg, France (http://vizier.u-strasbg.fr/)

2.2. SELECTION OF GALAXIES

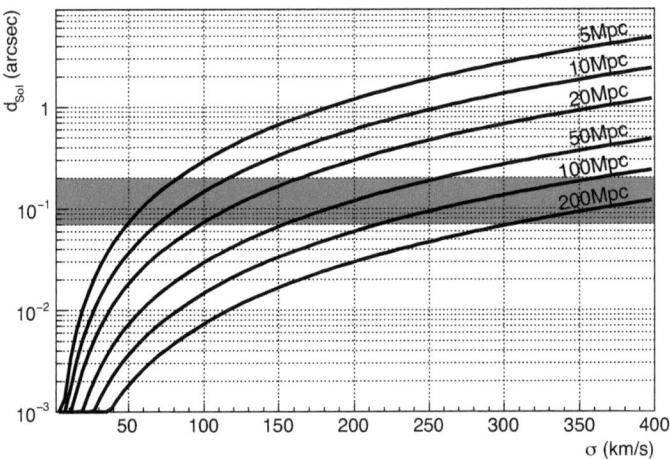

Figure 2.3: Diameter of the sphere of influence in arcseconds as a function of velocity dispersion σ for different distances, assuming that $M_\bullet$ follows the $M_\bullet$-σ relation of Tremaine et al. (2002). The red shaded region marks the spatial resolution range achieved with SINFONI in $\sim 75\%$ of the observing time, when the weather conditions were good or excellent.

as possible, in order to determine a precise stellar mass density. If the distance is overestimated, the mass of the SMBH would also be overestimated. The diameter of the SoI has to be resolvable with SINFONI under good to average seeing conditions, i.e. it has to be at least in the range $\sim 0.08'' - 0.2''$ (see Fig. 2.3). This limits the list of pseudobulge and low-σ objects to galaxies at distances $D \lesssim 20$ Mpc. The smaller the velocity dispersion of the galaxy the smaller the limiting distance out to which the SoI can be resolved. The SoI of high-σ galaxies ($\sigma > 300$ km s^{-1}) can be resolved out to at least $100-200$ Mpc, but unfortunately the spectral features used to determine the kinematics are redshifted out of the K-band with increasing distance. Already at 60 Mpc the first two CO bandheads are shifted to a region with enhanced sky absorption, such that this distance is imposed as an upper limit. The number density of these objects is very small, therefore also in this object category only very few objects are left. As many nearby late-type galaxies do not have a velocity dispersion measurement, we took major and minor axis longslit spectra of a number of those galaxies using the low-resolution spectrograph (LRS) at the Hobby-Eberly Telescope (HET) in Texas.

The inclination is also an important factor. The inclinations of galaxies close to face-on ($i \lesssim 40°$) usually cannot be determined very accurately, which inhibits the determination of the deprojected rotation velocity $v_{\mathrm{circ}} = v_{\mathrm{rot}}/\sin(i)$. For truly face-on galaxies, the measured rotation velocity is close to zero, thus no circular velocity can be derived. In truly edge-on galaxies the circular velocity can be directly measured and also the deprojection of the photometry has a unique solution, which makes dynamical modelling easier. It is not possible, however, to assess the presence or absence of a pseudobulge or classical bulge in the centre. Thus the selected pseudobulge objects have an inclination between 40° and 80°.

The remaining selection criteria are of more technical nature. The AO correction depends on the airmass, as the seeing increases with increasing airmass. The shorter the way of the galaxy light through the Earth's atmosphere, the better the AO correction, which is best when the object is in the zenith. The elevation limit of the VLT is 20°, but for a reasonable AO correction the object should be above $\sim 45°$ (corresponding to an airmass of ~ 1.4) for a few hours per night. This is particularly important for galaxies that are observed with the LGS, as the LGS AO correction depends stronger on the seeing than NGS AO correction. Therefore only objects within a declination range $-70° \lesssim \delta \lesssim 13°$ have been considered. In addition, due to the Earth's rotation around the Sun, each galaxy is visible at night only for some months per year, thus the observation time and the objects have been chosen carefully in order to be able to observe the selected galaxy sample during the few nights which were assigned to us by ESO per semester.

Most late-type galaxies, pseudobulges and core galaxies have a very low central surface brightness. It is not reasonable to spend an excessive amount of time on one galaxy, therefore the sample was restricted to objects that do not need more than $0.5 - 1$ night of exposure time (including overhead) to obtain a minimum S/N per pixel of around 40. This means effectively that the central K-band surface brightness should not be fainter than ~ 15 mag/arcsec^2. 2MASS (Skrutskie et al., 2006) provides near-IR surface brightness profiles for many nearby galaxies.

The last criteria (which turned out to be the most stringent ones) are the availability of the LGS and a TTS or the availability of an NGS. Whether the criteria for a TTS or NGS are fulfilled can be checked by measuring the R-band magnitude of the nucleus or a nearby star using available R-band imaging (e.g. from *HST* or SDSS). Many selected galaxies have a too faint nucleus to be observed in NGS mode, but are usually suitable as TTS. Unfortunately, during most of the observing time prior to the second commissioning in 2008, the LGS was unavailable.

2.3 Observations

The observations with SINFONI were carried out in visitor mode between March 2005 and May 2009. Table 2.2 gives a summary of the observations. The observation time allocated by ESO was

2.3. OBSERVATIONS

Table 2.1: Properties of the observed galaxies. Distance D and velocity dispersion σ are estimates based on literature values. d_{Sol} and $M_\bullet$ were calculated from these estimates using the $M_\bullet$-σ relation of Tremaine et al. (2002).

Galaxy	RA	Dec	Morph	type	D (Mpc)	σ (km s^{-1})	d_{Sol} ($''$)	$M_\bullet^{\text{est}}$ ($M_\odot$)	Nucleus
ESO138-G005	16h53m53.3s	-58d46m41s	SB0	high-σ	35	323	0.45	9.3×10^8	
NGC 307	00h56m32.6s	-01d46m19s	S0	high-σ	52.8	311	0.28	8.0×10^8	
NGC 1316	03h22m41.7s	-37d12m30s	SAB0	merger	18.6	228	0.42	2.3×10^8	FR I, L
NGC 1332	03h26m17.3s	-21d20m07s	S0	high-σ	22.9	321	0.68	9.0×10^8	
NGC 1374	03h35m16.6s	-35d13m35s	E3	core	19.8	185	0.26	9.9×10^7	
NGC 1398	03h38m52.1s	-26d20m16s	SBab	pseudo	18	200	0.33	1.3×10^8	S
NGC 1407	03h40m11.9s	-18d34m49s	E0	high-σ	28.8	273	0.39	4.7×10^8	
NGC 1550	04h19m37.9s	02d24m36s	SA0	high-σ	48.5	336	0.35	1.1×10^9	
NGC 3091	10h00m14.3s	-19d38m13s	E3	high-σ	49.2	319	0.31	8.8×10^8	
NGC 3351	10h43m57.7s	11d42m14s	SBb	pseudo	8.1	67	0.08	1.7×10^6	H
NGC 3368	10h46m45.7s	11d49m12s	SABab	pseudo	10.4	98.5	0.14	7.8×10^6	L2
NGC 3412	10h50m53.3s	13d24m44s	SB0	low-σ	11	101	0.14	8.7×10^6	
NGC 3489	11h00m18.6s	13d54m04s	SAB0	pseudo	12.1	91.1	0.10	5.7×10^6	T2/S2
NGC 3627	11h20m15.0s	12d59m30s	SABb	pseudo	10	115	0.20	1.5×10^7	T2/S2
NGC 3923	11h51m01.8s	-28d48m22s	E4	merger	21.1	255	0.46	3.6×10^8	
NGC 4371	12h24m55.4s	11d42m15s	SB0	pseudo	16.9	125	0.14	2.0×10^7	
NGC 4472	12h29m46.7s	08d00m02s	E2	core	17.1	294	0.76	6.3×10^8	S2::
NGC 4486A	12h30m57.7s	12d16m13s	E2	low-σ	16	110	0.11	1.2×10^7	
NGC 4486B	12h30m32.0s	12d29m25s	cE0	mid-σ	16	185	0.32	9.9×10^7	
NGC 4501	12h31m59.2s	14d25m14s	SAb	pseudo	13	161	0.30	5.6×10^7	S2
NGC 4536	12h34m27.0s	02d11m17s	SABbc	pseudo	25.4	182	0.19	9.2×10^7	H
NGC 4569	12h36m49.8s	13d09m46s	SABab	pseudo	16	117	0.13	1.6×10^7	T2
NGC 4579	12h37m43.5s	11d49m05s	SABb	pseudo	16	154	0.22	4.7×10^7	S1.9/L1.9
NGC 4699	12h49m02.2s	-08d38m53s	SABb	pseudo	19	215	0.36	1.8×10^8	S
NGC 4751	12h52m05.0s	-42d39m36s	SA0	core/high-σ	25.9	349	0.71	1.3×10^9	
NGC 4762	12h52m56.0s	11d13m51s	SB0	mid-σ	14.2	147	0.23	3.9×10^7	L2
NGC 5018	13h13m01.0s	-19d31m05s	E3	merger	40.8	211	0.23	1.7×10^8	
NGC 5102	13h21m57.6s	-36d37m49s	SA0	low-σ	4	65	0.15	1.5×10^6	
NGC 5328	13h52m53.3s	-28d29m22s	E1	high-σ	61.5	307	0.23	7.6×10^8	
NGC 5419	14h03m38.8s	-33d58m42s	E	high-σ	69.1	351	0.27	1.3×10^9	R
NGC 5516	14h15m54.7s	48d06m53s	E	high-σ	56.1	313	0.26	8.2×10^8	
NGC 7619	23h20m14.5s	08d12m23s	E	core/high-σ	53.0	322	0.30	9.2×10^8	

CHAPTER 2. OBSERVATIONS AND DATA REDUCTION

GTO (guaranteed time observations) time, i.e. time given to our group as a form of payment for being involved in building an instrument for an ESO telescope. The observations between March 2005 and March 2007 (i.e. the objects relevant for this thesis) and in March 2009 were carried out by myself.

Almost all observations were AO-supported except when the seeing was exceptionally bad. The LGS became operational in 2007, therefore all observations before were done in NGS mode. In 2007 it was partly possible to use the LGS, but it was rather unstable and sometimes the returned power was just not high enough to provide an acceptable AO correction during excellent seeing conditions. Only the galaxies NGC 3368 and NGC 3627 could be observed in LGS mode. After the observing run in March 2007 the LGS was unavailable until about October 2008. Due to the lack of NGS objects some nights of GTO time have been postponed to March 2009. Form November 2008 the LGS was stable and running, so whenever the weather conditions were good enough, exclusively LGS galaxies were observed.

The exposure time required to achieve a minimum central S/N of ~ 40 was determined beforehand by using 2MASS surface brightness profiles, S/N maps of already observed galaxies and the SINFONI exposure time calculator.[2] Depending on the atmospheric conditions and the surface brightness profile in the inner $1''$, which cannot be resolved by 2MASS, this value may be over- or underestimated. Therefore the S/N was always measured online after each observation block and the exposure time was adjusted if necessary. The S/N was measured by fitting a convolved stellar template spectrum to the CO bandhead region of a few spectra in the central region of the galaxy (details are given in Chapter 3). For this purpose the data were reduced quickly using a set of calibration files from previous observations. This is, of course, not very accurate but sufficient for a rough S/N estimate. From 2006 on, the ESO reduction pipeline was working and providing a quick data reduction online, such that already a few minutes after the observation block finished the data could be analysed.

The galaxies were observed in the K-band ($1.95 - 2.45$ μm) due to the better AO correction and spectral resolving power ($R = \lambda/\Delta\lambda \approx 5000$, see Section 3.4.2) compared to the J and the H-band. The 100mas scale was almost always used. Galaxies with a very small SoI were observed using the 25mas platescale. During bad seeing the 250mas scale was used. The observation time for a galaxy was split into blocks of one hour each, consisting of 10 min exposures of the sequence "object-sky-object-object-sky-object", thus for 40 min on-source exposure time 20 min of overhead was spent on the sky. Sky observations were required because in the near-infrared the sky emission of the atmosphere is strong and can only be satisfactorily removed with simultaneously taken sky

---

[2] http://www.eso.org/observing/etc/

2.3. OBSERVATIONS

exposures (see below). The sky region was selected based on existing imaging (*HST*, SDSS etc.) as an empty region as far away from the galaxy centre as possible along the minor axis, but still well within $\approx 2'$ in order to avoid losing the guide star. The movement of the telescope to the object position and doing the acquisition of the galaxy (in NGS mode) usually took about 10 – 15 min.

An integration time of 10 min per exposure was appropriate for most galaxies in the sample. Shorter integration times of, e.g., 5 min were chosen for very bright and extended objects as was done for NGC 1316 on the 250mas scale. Although even in this case the detector would not be saturated during longer exposure times, the so-called memory effect becomes significant when observing bright extended objects. This manifests in an artifact similar to an inadequate flat-fielding. The continuum-shape of the spectra is changed in a certain, relatively broad wavelength range – in some slitlets more, in others less – resulting in darker and brighter "stripes" in the object image. It is difficult to correct for this effect (see Ádámkovics et al. 2006), as no calibrations can be done for that purpose. The effect is small in the CO bandhead region, but becomes important when measuring the near-IR line indices which critically depend on the continuum shape. Also the photometry of such a striped image is unusable.

Longer integration times (15 min) can be chosen for faint targets. As the sky subtraction becomes more difficult the longer the integration time is, only 10 min were used by default for each galaxy.

The exposures within an observation block were slightly spatially shifted with respect to each other. This has two reasons. The detector has quite a large number of bad pixels. These are removed using a bad pixel mask, but this might not cover all bad pixels or the correction might be suboptimal in regions with clusters of bad pixels. An additional method to remove the remaining bad pixels is to average several exposures and exclude those pixels that deviate by a certain factor from the others at the same spatial and spectral position. This requires that spectra from the same spatial position in the galaxy are recorded on different positions of the detector. This can be realized by spatial dithering, i.e. applying a small shift of a few spaxels in different spatial directions from exposure to exposure, as illustrated in Fig. 2.4. Another reason for spatial dithering is the rectangular shape of the spaxels (e.g. $0.05'' \times 0.1''$ for the 100mas scale). If no shift were applied, the spatial resolution in the y-direction would be smaller than the spatial resolution in the x-direction. Therefore half of the exposures per observation block were shifted by an odd number of half spaxels in y-direction, resulting in square spaxels of e.g. $0.05'' \times 0.05''$ size.

Most calibration exposures like darks and flatfields were taken during daytime by the Paranal Observatory staff. The only night-time calibrations needed for the data reduction and analysis are observations of the sky, telluric standard stars, PSF stars and kinematic template stars.

CHAPTER 2. OBSERVATIONS AND DATA REDUCTION

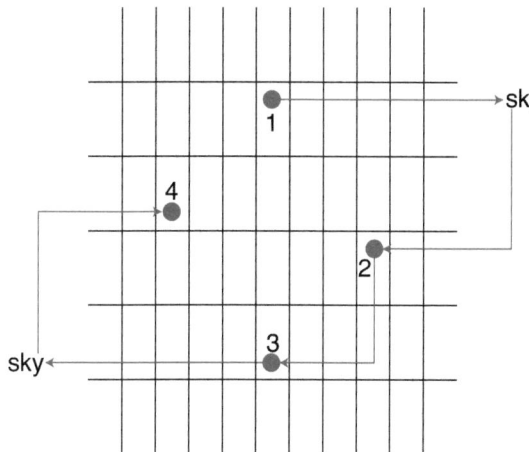

Figure 2.4: Illustration of the spatial dither pattern for one "object-sky-object-object-sky-object" observation block. The grid represents the rectangular spaxels in part of the SINFONI field of view. The four red points are the "object" positions of the centre of the observed object.

Near-IR airglow emission

In the near-IR the night-sky is very bright and the dominant source of background. The airglow emission in the K-band comes from the excited hydroxyl (OH*) molecule in the mesosphere and has a maximum at an altitude of ~ 87 km. OH radicals are created in the reaction $H+O_3 \rightarrow OH+O_2 + 3.34$ eV. This results in a system of 45 rotation-vibration bands between 0.38 and 4.5 µm due to transitions among the nine lowest levels of the ground electronic state of the hydroxyl radical. They are also called the Meinel bands (Meinel, 1950). They are strongest in the near-IR and their total intensity is ~ 18000 times higher than that of the strongest night-sky emission line in the optical, atomic oxygen at $\lambda = 5577$ Å. Each band has a complex structure and its intensity shows temporal and spatial variability due to changes in the distribution and production rate of ozone, and the local temperature and density. For near-IR observations the short-term variability of the airglow emission is of major concern. This emission is caused by the passage of atmospheric waves ("gravity waves") with periods $\geqslant 5$ min through the region where the OH emission is generated. The amplitude of these gravity waves increases with decreasing atmospheric density. When the amplitude becomes too large, these waves break and transfer their energy and momentum into the atmosphere. The resulting airglow variations have amplitudes of typically $\sim 10\%$. Thus ideally, in order to subtract the night-sky emission from the object exposures, a simultaneous observation of the night sky is needed. In the case of longslit spectroscopy, the slit is usually more extended than the object, so there is no need for an additional sky exposure. SINFONI, however, like most integral field instruments, has a small field of view, which is completely

2.3. OBSERVATIONS

Table 2.2: Observation log of the galaxies observed with SINFONI and the achieved spectral resolution.

Galaxy	date	platescale	t_{exp} (min)	PSF (″)	AO mode
ESO138-G005	2009-05-18	250mas	40	0.36	NGS
NGC 307	2008-11-26	100mas	40	0.23	LGS
	2008-11-27	100mas	80	0.34	LGS
NGC 1316	2005-10-10	250mas	30	0.48	no-AO
	2005-10-12	100mas	70	0.12	NGS
	2005-10-11	25mas	100	0.09	NGS
	2005-10-12	25mas	40	0.09	NGS
NGC 1332	2008-11-25	250mas	40	0.63	no-AO
	2008-11-25	100mas	80	0.16	NGS
NGC 1374	2008-11-28	100mas	80	0.13,0.34	LGS
NGC 1398	2006-09-16	100mas	100	0.19	NGS
	2006-09-18	100mas	110	0.32	NGS
	2008-11-24	100mas	40	0.15	NGS
	2008-11-25	100mas	40	0.14	NGS
NGC 1407	2008-11-24	100mas	200	0.20	LGS
NGC 1550	2008-11-27	100mas	120	0.17,0.35	LGS
	2008-11-28	100mas	20		LGS
NGC 3091	2008-11-26	250mas	40		no-AO
	2008-11-27	250mas	10		no-AO
	2008-11-28	250mas	40		no-AO
	2008-11-26	100mas	80	0.15	NGS
	2009-04-19	100mas	40	0.13	NGS
	2009-04-20	100mas	40	0.13	NGS
	2009-04-22	100mas	40	0.13	NGS
NGC 3351	2009-04-19	100mas	40	0.20	LGS
	2009-04-20	100mas	40	0.18	LGS
	2009-04-21	250mas	40		no-AO
NGC 3368	2007-03-23	100mas	80	0.16	LGS
	2007-03-24	100mas	60	0.23	LGS
NGC 3412	2008-03-07	100mas	80	0.13	NGS
	2008-03-08	100mas	40	0.13	NGS
	2008-03-11	25mas	10	0.13	NGS
NGC 3489	2007-03-22	25mas	40	0.08	NGS
	2007-03-24	25mas	80	0.08	NGS
NGC 3627	2007-03-25	250mas	80	0.74	no-AO
	2007-03-21	100mas	80	0.24	LGS
	2007-03-25	100mas	50	0.27	LGS
	2007-03-22	25mas	60	0.087	LGS

Continued on next page

Table 2.2 – continued from previous page

Galaxy	date	platescale	t_{exp} (min)	PSF (″)	AO mode
	2009-04-22	250mas	20		no-AO
	2009-05-18	250mas	20		no-AO
NGC 3923	2009-05-18	250mas	80	0.33,0.36	NGS
NGC 4371	2009-03-20	100mas	80	0.14	LGS
	2009-03-21	100mas	40	0.25	LGS
	2009-03-22	100mas	40	0.20	LGS
NGC 4472	2009-04-20	100mas	40	0.14	NGS
	2009-04-21	100mas	40	0.14	NGS
	2009-04-23	250mas	60	0.33	NGS
NGC 4486A	2005-04-06	100mas	80	0.11	NGS
	2005-04-07	100mas	60	0.09	NGS
	2005-03-22	25mas	80	0.07	NGS
	2005-03-22	25mas	40	0.07	NGS
NGC 4486B	2009-05-18	100mas	40	0.16	NGS
	2009-05-19	100mas	40	0.14	NGS
	2009-05-22	100mas	20	0.17	NGS
	2009-05-22	250mas	40	0.65	no-AO
NGC 4501	2008-03-12	100mas	80	0.13	NGS
NGC 4536	2009-04-19	100mas	60	0.18	LGS
	2009-04-21	250mas	60		no-AO
	2009-04-22	100mas	80	0.18	NGS
NGC 4569	2008-03-08	100mas	80	0.15	NGS
NGC 4579	2008-03-08	100mas	40	0.15	NGS
	2008-03-09	100mas	80	0.15	NGS
NGC 4699	2008-03-10	100mas	80	0.13	NGS
	2008-03-11	100mas	40	0.13	NGS
NGC 4751	2008-03-07	250mas	40		no-AO
	2008-03-12	250mas	20		no-AO
	2009-03-19	100mas	40	0.14	LGS
	2009-03-21	100mas	40	0.15	LGS
NGC 4762	2008-03-09	100mas	40	0.30	NGS
	2008-03-12	100mas	50	0.13	NGS
NGC 5018	2009-05-20	100mas	40	0.15	NGS
	2009-05-20	100mas	40	0.18	NGS
	2009-05-20	250mas	20	0.60	no-AO
	2009-05-21	100mas	40	0.14	NGS
NGC 5102	2007-03-21	100mas	80	0.14	NGS
	2007-03-23	100mas	40	0.16	NGS
	2007-03-22	25mas	60	0.086	NGS

Continued on next page

2.3. OBSERVATIONS

Table 2.2 – continued from previous page

Galaxy	date	platescale	t_{exp} (min)	PSF (″)	AO mode
	2007-03-24	25mas	20	0.068	NGS
NGC 5328	2008-03-11	100mas	80	0.17	NGS
	2009-03-19	100mas	40		NGS
	2009-04-22	250mas	80		no-AO
	2009-04-23	100mas	40	0.13	NGS
	2009-04-24	100mas	90	0.12	NGS
NGC 5419	2009-05-19	100mas	10		LGS
	2009-05-20	100mas	10		LGS
	2009-05-21	100mas	100	0.17	LGS
	2009-05-22	100mas	60	0.17	LGS
	2009-05-23	100mas	150	0.17,0.21	LGS
NGC 5516	2009-03-20	100mas	80	0.14	LGS
	2009-03-21	100mas	20	0.12	LGS
	2009-03-22	100mas	100	0.2	LGS
NGC 7619	2008-11-24	100mas	40	0.3	LGS
	2008-11-26	100mas	40	0.19	LGS
	2008-11-28	100mas	40	0.17	LGS

Two values for the PSF are given when it was measured twice during one night, e.g. due to a very long exposure time or changing weather conditions.
For galaxies with unreduced data or no measured PSF star no value for the PSF is given.

filled with light from the object. Therefore, before or after each object exposure a blank sky region was observed, which is sufficiently far away to be not affected anymore by the galaxy light and close enough to be not significantly affected by spatial variations of the sky emission. The selected sky regions were 1-2′ away from the galaxy centres. As an integration time of 10 min per exposure was chosen, significant flux variations of the OH bands may be present between the object exposure and the corresponding sky exposure, which would result in an incorrect sky subtraction. Even worse, instrumental flexure due to the movement of the instrument with time can produce small shifts of the wavelength scale, complicating the sky subtraction even more (see Section 2.4).

Telluric standards

The galaxy light has to pass the Earth's atmosphere and thereby interacts with molecules like water vapour and oxygen, which results in telluric absorption lines in the spectrum of the galaxy. The transmissivity of the atmosphere is significantly lower in the near-IR than in the optical and varies with time and airmass. In order to remove these strong telluric features, a star with a known or featureless spectrum has to be observed shortly before or after the observation of the galaxy. This star should not be too far away from the galaxy and be at a zenith distance that corresponds to the average zenith distance of the galaxy during the observations. Mid to late B-type dwarfs from the Hipparcos catalogue with a known K-band magnitude were observed. The advantage of stars of this type is that their spectra are almost featureless apart from hydrogen recombination lines. In the K-band the only absorption feature is Brγ at 2.166 µm, which can be easily removed. Stars of later type would have some CO absorption and other absorption lines. One object exposure and one sky exposure of the telluric stars were sufficient, as just a 1-dimensional spectrum – a combination of all spectra within a certain aperture – is used for the telluric correction. The stars are usually very bright ($R \approx 6 - 8$ mag), therefore the exposure time is only of the order of a few seconds.

The point-spread function

In order to measure the achieved spatial resolution, i.e. the FWHM of the point-spread function (PSF), and the Strehl ratio, a "PSF star" was observed regularly (at least once per object and night). For NGS observations, the NGS itself was used as PSF star in case it was a star and not the nucleus of a galaxy. This situation was given only for NGC 4486a. Due to the effect of anisoplanatism one should ideally observe a star with the same R-band magnitude and $B - R$ colour as the inner 3″ of the galaxy using another star with the magnitude and colour of the NGS star at the same distance as the galaxy-NGS separation. As the NGS of NGC 4486a is only $\sim 2″$ away from the nucleus the effect of anisoplanatism is weak and therefore the NGS was directly observed as a PSF star. This could be done within the same observation block, such that no time was lost due to repeated acquisitions. For the other galaxies in Table 2.1 a nearby star with the same R-band magnitude and $B - R$ colour as the central 3″ of the galaxy nucleus was selected and observed using the same setup as for the galaxy observation (i.e. same band and platescale using the target itself for AO correction). The integration time was chosen depending on the brightness of the star and is usually of the order of a few minutes. For LGS observations the procedure was similar. The galaxy nucleus was used as TTS, thus a PSF star with the same magnitude and colour was selected.

There are several problems associated with the measurement of the spatial resolution via PSF star observations. First of all it is very inefficient to frequently stop the galaxy observations in order to observe a PSF star, as it takes (depending on whether NGS or LGS is used) each time 15 – 30 min. Thus a PSF star can effectively be observed only once per object and night. The most severe problem is therefore the time offset between object and PSF star observations. The atmospheric conditions are not constant in time, and in several cases it happened that the seeing during the PSF star observations was significantly different than the average seeing during the galaxy observations. In addition the response of the wavefront sensor on a star is different from

2.3. OBSERVATIONS

the response on an extended object like a galaxy nucleus. This is usually not a severe problem for strong AGN with pointlike unresolved nuclei, but for inactive galaxies this adds more uncertainties to the PSF measurements. These uncertainties might be particularly high for the core galaxies, as the cores have a more or less flat surface brightness profile out to radii of several arcseconds. PSF star observations are presently the only method to get an estimate on the PSF for inactive galaxies. An ideal alternative would be the reconstruction of the PSF from the wavefront sensor data. This method has been recently implemented for NACO, a near-infrared imager which is also installed at the VLT UT4, but it is not yet available for SINFONI. Thus the only solution is to use the PSF star measurements and, in cases with reasonable doubts about the reliability of the PSF, to investigate possible deviations from the true PSF by comparing the SINFONI image of the galaxy with a higher-resolution image in the same band with a known PSF. This is also difficult, as K-band images are rarely available with a similar or higher resolution than that of SINFONI. For NGC 1316 the 100mas SINFONI image was therefore compared with the 25mas SINFONI image, and in the case of NGC 3368 the SINFONI image was compared with a dust-corrected H-band *HST* NICMOS2 image (see Section 6.3). For galaxies with a strong, unresolved AGN it is possible to measure the PSF from the data itself (from the AGN core, the change of the continuum slope, the CO bandhead equivalent width or the Brγ emission from the broad line region, see e.g. Davies 2008; Davies et al. 2006, 2004; Mueller Sánchez et al. 2006). The observed galaxies, however, are at most weakly active and therefore not suitable for this kind of analysis.

Stellar kinematic templates

For the measurement of the stellar kinematics (see Chapter 3) a number of kinematic template stars (K or M stars) were observed during bad weather conditions. Bright stars with $K \sim 5 \pm 2$ mag were selected, as a high S/N per pixel within a short exposure time is desired. Special care was taken not to saturate the detector with such bright stars, therefore an NDIT > 1 (NDIT = number of detector integration times) was chosen and the AO correction was reduced when necessary. Sometimes the weather conditions were so bad that clouds reduced the brightness of the template stars.

The kinematic template stars were observed in blocks of 3 – 4 "object-sky" cycles, and the object positions were shifted randomly around a fixed position. As the exposure times were very short, spectral dithering was turned on. None of the galaxies were observed with spectral dithering. Should this be required in the future it will not be necessary to observe the templates again. The template observations were scheduled on all three platescales, but as the observation of the galaxies has priority the template observations were interrupted after finishing a platescale in case the weather conditions improved. The S/N of the telluric standard is critical for the S/N of the respective kinematic template. Therefore a longer integration time for the tellurics was needed.

2.4 Data reduction

The data reduction methodology of IFS data is basically identical to the data reduction of longslit data, with some additional steps needed to reconstruct the 3D datacube. The software used for the data reduction are SPRED (Abuter et al., 2006; Schreiber et al., 2004), IDL,[3] QFitsView[4] and the ESO pipeline (Modigliani, 2009). SPRED is a pipeline developed at MPE for the reduction of SINFONI data. It is written in ANSI C, makes use of the C library eclipse for handling fits files and can be employed by the user via a collection of PYTHON scripts. It was available already during commissioning of SINFONI in 2004. The software package and some commissioning data (for training purposes) were kindly provided by F. Eisenhauer. SPRED was solely used for the data reduction of the 2005 data. After the commissioning of SINFONI, ESO started to develop a reduction pipeline for common use with the front-ends Gasgano and EsoRex, in the same way as the pipelines of other ESO instruments. This pipeline is based on SPRED, thus the basic structure is the same with just a few small differences. Both pipelines are actively supported and regularly improved. In addition to SPRED, several IDL programs have been developed by R. Davies that improve shortcomings encountered with the SINFONI data, e.g. deliver a better sky subtraction (Davies, 2007) or bad pixel correction. The improved sky subtraction program has recently been implemented in the ESO pipeline. As it is possible to use SPRED and the ESO pipeline in parallel, for each step the pipeline that yielded better results was chosen to reduce the data from the 2006+ runs. Initially the majority of steps were done with SPRED, lately mainly the ESO pipeline was used. A detailed description of the data reduction steps and the recipes used by the ESO pipeline can be found in the ESO SINFONI pipeline user manual (Modigliani, 2009).

QFitsView is a program for displaying and manipulating datacubes. It was used throughout the data reduction procedure and beyond for the inspection of intermediate results and the final object datacubes. The most common features can be used interactively. Being based on DPUSER,[5] a large number of commands can be used and scripts can be executed.

In the remaining chapter the basic data reduction steps performed on the SINFONI data are described.

---

[3] http://www.ittvis.com/
[4] http://www.mpe.mpg.de/~ott/QFitsView/
[5] http://www.mpe.mpg.de/~ott/dpuser/

2.4. DATA REDUCTION

Bad lines removal

The bias level of each line in the detector is estimated from the four unilluminated pixels at the rim of the detector. If one or more bad pixels are present in these eight pixels per line (four pixels on each end), the bias level for this line is overestimated. The measured bias is subtracted from each line during the data processing at detector level, resulting in raw frames with some dark stripes. The effect increases with exposure time. It has to be corrected before starting the data reduction. A DPUSER script kindly provided by S. Gillessen and later an IDL script provided by ESO were used for this purpose.

Dark frames

Dark frames are necessary to identify bad pixels and they need to be subtracted from the science exposures in case the corresponding sky frame is not subtracted. For each object exposure time three dark exposures with the same integration time were taken during the day-time calibrations by ESO. These three exposures were combined to a single exposure, thereby eliminating possible cosmic rays.

Flatfields

For each combination of grating and platescale used for the science observations a set of five exposures to the light of a halogen lamp and five lamp-off exposures were done during the daytime calibrations. The lamp-off frames were subtracted from the lamp-on frames and the resulting files were combined.

Bad pixel masks

The dark frames and the flatfield were used to generate bad pixel maps, which were then combined to a master bad pixel map. Generating bad pixel maps is not trivial in particular for the 25mas scale, where often the slitlet edges are marked as bad. This can perturb the wavelength calibration. The number of the pixels flagged as bad were adjusted by changing certain parameters like the detection threshold. The final master bad pixel maps had around 1×10^5 bad pixels from $\sim 4.2 \times 10^6$ pixels in total, i.e. $\sim 2.4\%$. This seems a lot, but the majority of bad pixels consisted of the outermost columns and rows not illuminated, a very large cluster and a few small clusters of bad pixels and a few bad columns, thus apart from these regions the fraction of bad pixels was only about 1%.

CHAPTER 2. OBSERVATIONS AND DATA REDUCTION

Detector distortion

The curvature of the detector was determined by recording a series of spectra from a continuous light fibre, which was moved perpendicular to the slitlets. This was also done during the daytime calibrations for all gratings used during the night. The resulting ~ 75 spectra were co-added and the distortion was computed by tracing the fibre spectra and solving a 2D polynomial fit which transforms the coordinates of the distorted frame to undistorted coordinates. The relative distance (in pixels) between the slitlets was determined simultaneously.

Wavelength calibration

The wavelength calibration files are spectra of Neon and Argon arc lamps, taken during daytime for each combination of grating and platescale used during the night. The spectrum was background subtracted, flatfielded, corrected for bad pixels and distortion. The Ne and Ar lines were identified by crosscorrelating the spectrum with a spectrum made from a reference line list and convolved with a Gaussian. A polynomial was then fitted along each column to determine the dispersion coefficients, which were smoothed and used to generate a wavelength calibration map. At the same time the edges of the slitlets were determined from the positions of the brightest arc emission lines.

Telluric correction

A telluric standard star was always observed directly before, after or in between the galaxy observations. The telluric frames were sky subtracted, flatfielded, corrected for distortion and bad pixels and wavelength calibrated. Due to the short exposure time a simple subtraction of the sky frame worked well without leaving sky residuals. The datacube was reconstructed and a 1D standard star spectrum was extracted by averaging all spectra within a certain aperture. As the spectra of mid-to-late type B stars are featureless except for the Brγ absorption line, this line was removed in all telluric spectra using the task splot in IRAF, by fitting a Voigt profile to the line and subtracting the fit. Active or starforming galaxies often show Brγ emission, but this line is usually redshifted to wavelengths which are not affected by the Brγ absorption of the telluric. In order to remove the continuum shape of the star, the spectrum of the telluric was divided by a blackbody spectrum of a temperature that corresponds to the spectral type of the star. The final result, after continuum normalisation, is the transmission of the atmosphere as a function of wavelength (see Fig. 2.5).

2.4. DATA REDUCTION

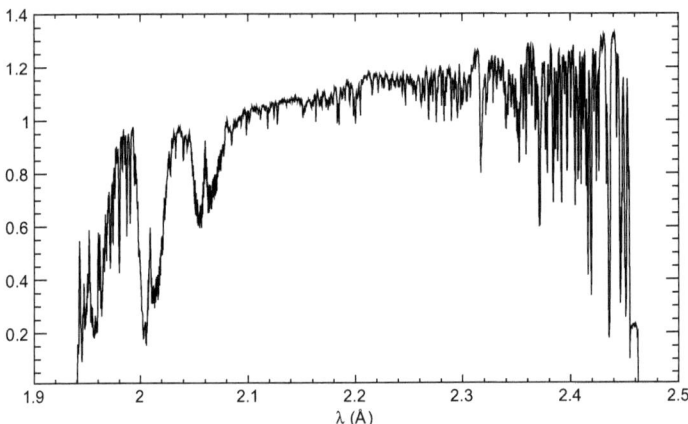

Figure 2.5: Atmospheric transmission as a function of wavelength created from a B9V star spectrum.

Spatial resolution and Strehl ratio

The PSF star exposures were calibrated in the same way as the telluric star. The Strehl ratio was computed as a function of wavelength by comparing the PSF star image with a theoretical PSF for a telescope with the primary and secondary mirror diameters of the VLT. The combined cube was collapsed along the wavelength direction (i.e., at each spatial position, the intensity was integrated within a certain wavelength range). The spatial resolution, i.e., the FWHM of the image of the star, was measured by fitting a 2D Gaussian to the image (these values are given in Table 2.2). For the dynamical modelling, however, the normalised 2D image of the PSF star was used instead of a 2D fit.

Reconstruction of the object datacubes

The galaxy exposures were, like the PSF, sky subtracted, flatfielded, corrected for distortion and bad pixels and wavelength calibrated. The bad pixels were corrected by using a 3D Bezier interpolation. Then the datacubes were reconstructed using the relative slitlet positions. The telluric absorption lines were removed by dividing each spectrum in the cube by the normalised transmission spectrum of the atmosphere. The final step was the combination of all object datacubes. Fig. 2.6 shows the final 3D datacube for the NGC 4486a (cf. Chapter 4).

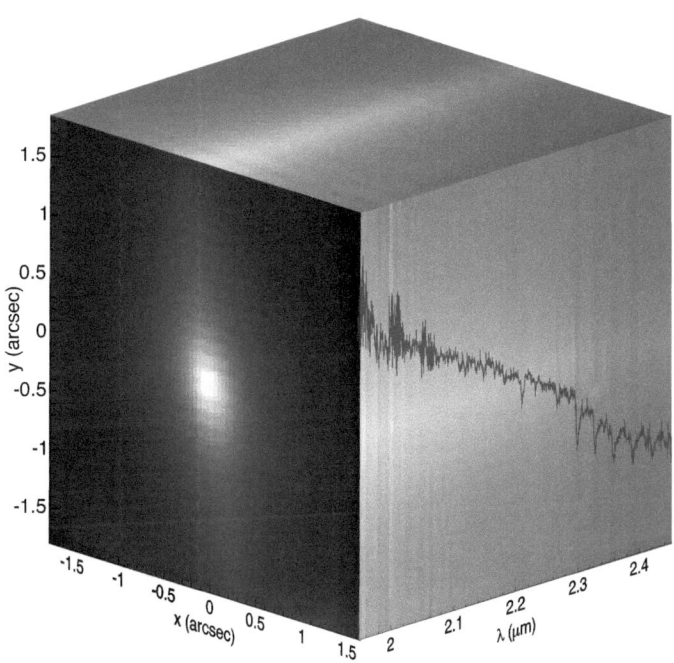

Figure 2.6: 3D illustration of the final datacube of NGC 4486a. A 1D spectrum is overlaid along the λ direction for a better orientation.

2.4. DATA REDUCTION

Improvement of the sky subtraction and the wavelength calibration

The individual datacubes were examined for residual sky lines using QFitsView. In many cases significant residual sky lines or even "P-Cygni" shaped residuals were present in the spectra. "P-Cygni" residuals are the result of instrumental flexure due to the movement of the instrument with time, which causes a small shift of the wavelength scale. Thus when a sky spectrum with such a shifted wavelength scale is subtracted from an unshifted object spectrum, each skyline is insufficiently subtracted on one side, and overcorrected on the other side. A significant improvement of the sky subtraction could be achieved using the method of Davies (2007). With this technique the sky emission lines were scaled as a function of wavelength such that the sky background was optimally matched. For this purpose separate object and sky datacubes were created, which were calibrated in the usual way but instead of subtracting the sky background from the object, only darks were subtracted from both object and sky. A blackbody function was fitted to the thermal background of the sky and subtracted from object and sky cube. The positions of the strongest OH lines were calculated in both object and sky cube, from which the shift in wavelength was determined and used to correct the wavelength scale of the sky spectrum. This step was omitted when no P-Cygni residuals were present. Nevertheless the shifts between the object cubes, in particular cubes from different nights, were determined and, if necessary, corrected with respect to one certain sky exposure. To derive a scaling function the code of Davies (2007) first divided the sky spectrum in regions with emission lines from transitions between vibrational bands, as each group of lines from a certain vibrational transition is to a good approximation located in a restricted wavelength region. A scaling function was determined that minimized the differences between object and sky. In a second step the same was done for certain rotational transitions. The emission lines from rotational levels could not be clearly identified and distinguished, thus only the most isolated lines were used to determine a scaling function. This was repeated for the rest of the spectrum, using the remaining isolated lines. The resulting scaling functions were combined and a corrected sky cube was created which was then subtracted from the object cube. With this method the sky subtraction could be significantly improved.

Flux calibration

A flux calibration was necessary in a few cases (NGC 1316 and NGC 3368) in order to measure the H_2 emission line fluxes. For NGC 1316 an already flux-calibrated K-band image with subarcsecond resolution was available, thus the flux within an $8'' \times 8''$ aperture was measured and applied to the 250mas SINFONI datacube. The 25mas and 100mas cubes were in turn calibrated using the 250mas cube (cf. Section 5.2). For the other galaxies analysed in this thesis no K-band image of sufficiently high spatial resolution was available, therefore the associated telluric standard stars which have measured 2MASS K-band magnitudes were used as a reference. A total spectrum of the star was extracted, the counts per second were measured and using the SINFONI K-band zero point magnitude of 4.14×10^{-10} W m^{-2} μm^{-1} the flux density for the magnitude of the telluric star was calculated. With this value a scaling was determined and applied to the object data cube, that converts counts/second to units of flux density (W m^{-2} μm^{-1}).

3
Stellar kinematics

The basic ingredient needed for the stellar dynamical modelling of the observed galaxies is a reliable measurement of the stellar motions using the absorption lines in the K-band galaxy spectra. This chapter first gives an overview of the available techniques developed to measure absorption-line kinematics and of the absorption features in the K-band. Then the suitability of the two most general techniques for K-band SINFONI spectra is analysed based on extensive simulations. At the end the detailed recipe used in Chapters 4-6 for the derivation of the stellar kinematics is given.

3.1 Extraction techniques

While for the Milky Way a direct measurement of the motions of individual stars is possible (e.g. Gillessen et al. 2009), no single stars can be resolved in the centres of other galaxies. Determining the stellar kinematics of these galaxies therefore is restricted to the velocity distribution of a large number of unresolved stars along the line of sight. A galaxy spectrum can be described as the convolution of the spectrum of the stars $S(\lambda)$ (assuming all stars have the same spectrum) with a broadening function, the so-called line-of-sight velocity distribution (LOSVD) $\mathscr{L}(v)$:

3.1. EXTRACTION TECHNIQUES

$$G(\ln \lambda) = S(\ln \lambda) \otimes \mathscr{L}(v). \tag{3.1}$$

Many techniques to reconstruct the LOSVDs from galaxy spectra have been developed (see review below), all of them require the use of one or several stellar templates – spectra of nearby stars of similar type as the stars in the galaxy, which contribute most to the luminosity. Early methods extract the LOSVDs in Fourier space. This is a reasonable approach as a convolution in real space is simply a multiplication in Fourier space, thus the LOSVD can be recovered easily without large demands regarding computing power. More recent methods fit the spectra directly in pixel space, which has the advantage that parts of the spectrum (e.g. emission lines and bad pixels) can easily be excluded. All methods can be further classified with respect to whether a non-parametric LOSVD is derived or an a priori assumption on the LOSVD shape is made. A Gaussian-shaped LOSVD is certainly a reasonably good first approximation in most cases, however, there is no physical reason why stars in a galaxy should have a purely Gaussian velocity distribution. Real LOSVDs indeed always show deviations from a Gaussian, which provide important information on the orbital structure and therefore are essential for the derivation of the mass of a central black hole. Most parametric techniques therefore use the Gauss-Hermite parametrization by van der Marel & Franx (1993) and Gerhard (1993):

$$\mathscr{L}(v) = \frac{\gamma \alpha(y)}{\sigma} \sum_{i=0}^{N} h_i H_i(y), \tag{3.2}$$

$$\text{with} \quad y = \frac{v - v_0}{\sigma} \quad \text{and} \quad \alpha(y) = \frac{1}{\sqrt{2\pi}} e^{-\frac{y^2}{2}},$$

where γ is the line strength, v_0 the recession velocity, v and σ the measured velocity and velocity dispersion, and $H_i(y)$ are the Hermite polynomials. The Hermite coefficients h_i describe the deviations from a Gaussian. If the LOSVD does not deviate too strongly from a Gaussian shape it is usually sufficient to truncate $\mathscr{L}(v)$ at $i = 4$:

$$\mathscr{L}(v) = \frac{\gamma \alpha(y)}{\sigma} [1 + h_3 H_3(y) + h_4 H_4(y)] \tag{3.3}$$

with

$$H_3(y) = \frac{1}{\sqrt{6}}\left(2\sqrt{2}y^3 - 3\sqrt{2}y\right) \qquad (3.4)$$

$$H_4(y) = \frac{1}{\sqrt{24}}\left(4y^4 - 12y^2 + 3\right). \qquad (3.5)$$

This results in four free parameters that describe the LOSVD: the velocity v, the velocity dispersion σ, and the higher-order moments h_3 and h_4. h_3 describes asymmetric deviations from a Gaussian profile (skewness). Symmetric deviations (kurtosis) are described by the parameter h_4. Some example LOSVDs are shown in Fig. 3.1.

In the following a selection of LOSVD extraction techniques are presented in chronological order. Methods used later in this work are explained in greater detail.

- The **Fourier Quotient** method (FQ, Sargent et al. 1977) is a parametric method which fits a Gaussian LOSVD to the quotient of the galaxy spectrum and the template spectrum in Fourier space. This method has several disadvantages, e.g. it is very sensitive to template mismatch (i.e. when the spectral type of the template does not match that of the galaxy spectrum), and the errors are strongly correlated and thus complicated to estimate. Later methods based on FQ but with a different error analysis are, e.g., the Fourier Difference (Dressler, 1979) and the Fourier Fitting (Franx et al., 1989) method.

- In the **Cross-Correlation** method (CC, Simkin 1974; Tonry & Davis 1979), the galaxy spectrum is cross-correlated against a stellar template spectrum. The peak of the galaxy-template correlation function is fitted by a Gaussian. The position of the peak gives the velocity and the velocity dispersion can be derived from the peak width. CC is less affected by template mismatch than FQ and the error estimation is easier.

- The **Fourier Correlation Quotient** method (FCQ, Bender 1990) is an improvement of the previous methods. It has the advantages that it is less sensitive to template mismatch than FQ and it is non-parametric, i.e. it recovers the full broadening function without a priori assumptions on the shape. This hybrid approach is based on the deconvolution of the peak of the template-galaxy correlation function with the peak of the autocorrelation function of the template. In contrast to the previous methods, where the complete correlation functions are deconvolved, deconvolving only the peaks is what makes this method so insusceptible to template mismatch. High-frequency components of the template-galaxy correlation function, caused by noise, would be strongly amplified during deconvolution. Therefore a Wiener

3.1. EXTRACTION TECHNIQUES

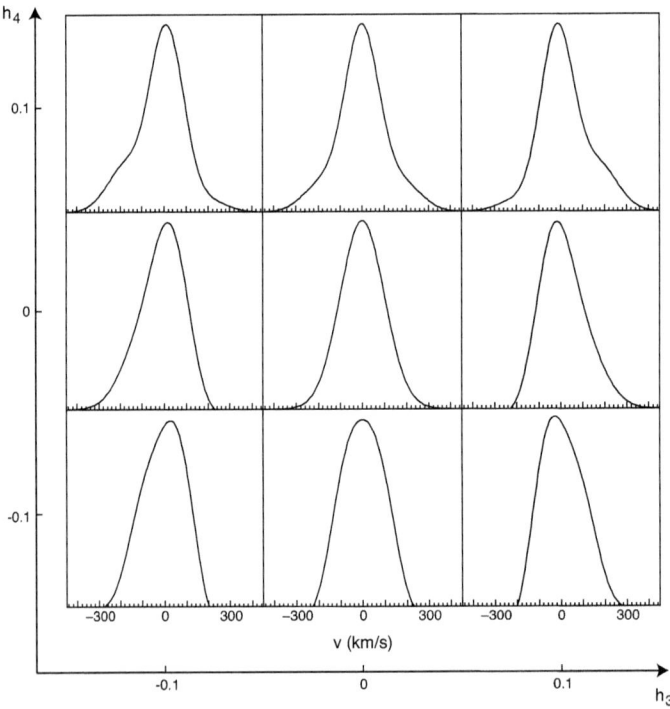

Figure 3.1: Illustration how the LOSVD shape changes with varying h_3 and h_4. The central LOSVD is purely Gaussian ($h_3 = h_4 = 0$). For $h_3 \neq 0$ the LOSVDs are skewed and for $h_4 \neq 0$ the LOSVDs are more peaked or more flattened.

filter is implemented which suppresses these components. This filter is obtained by fitting the template-galaxy correlation function with a Gaussian, using the fitting parameters to construct a model of its Fourier transform, and add a model for the noise. A smoothing parameter W determines the width of the Gaussian. $W = 1$ corresponds to "optimal" filtering, higher values produce less smoothing.

- Another non-parametric method is **Direct Fitting** (Rix & White, 1992) in pixel space, where the galaxy and template spectra are assumed to consist of several components (continuum, noise, absorption lines) and the best-fitting combination of components is determined by a least-squares minimisation.

- Winsall & Freeman (1993) developed a **Modified Fourier Quotient** technique which is able to deal with non-Gaussian LOSVDs. This method is parametric and works in Fourier space.

- The **Unresolved Gaussian Decomposition** (Kuijken & Merrifield, 1993) parametrizes the broadening function as a sum of pure Gaussians uniformly spaced in velocity.

- In the **Gauss-Hermite fitting method** a Gauss-Hermite expansion is used to parametrize the LOSVD and a least-squares minimisation is performed either in Fourier space (van der Marel & Franx, 1993) or in pixel space (van der Marel et al., 1994).

- Saha & Williams (1994) developed a **Bayesian Method** to extract the broadening function. It is able to cope with low-SN data and template mismatching, can be used either in a parametric or a non-parametric mode and the fitting is done in pixel space.

- The **Maximum Penalized Likelihood** method of Merritt (1997) is non-parametric and works in pixel-space. Here the broadening function is found by minimizing the penalized log likelihood

$$-\log \mathcal{L}_p = \sum_i [G(\lambda_i) - (\mathcal{L} \otimes S)_i]^2 + \alpha \mathcal{P}, \qquad (3.6)$$

where α is the smoothing parameter and $\mathcal{P}$ is a penalty function, which is large for noisy velocity distributions and zero for Gaussian ones. This method can also be applied to low-S/N data as even a large degree of smoothing does not bias the solutions too much.

- The likewise non-parametric **Maximum Penalized Likelihood** method (MPL, Gebhardt et al. 2000a) is based on Saha & Williams (1994) and Merritt (1997), but the LOSVDs and the weights of the template stars are simultaneously fitted. Both the LOSVDs and the template

3.2. SPECTRAL FEATURES

weights are constrained to be non-negative at all points. An initial binned velocity profile is convolved with a linear combination of the template spectra and the residuals of the resulting spectrum to the observed galaxy spectrum are calculated. The velocity profile and the weights of the templates are iteratively adjusted in order to provide the best match to the observed galaxy spectrum, which is derived by minimizing the function

$$\chi_P^2 = \chi^2 + \alpha \mathscr{P} = \chi^2 + \alpha \int_V \left(\frac{\partial^2 \mathscr{L}(v)}{\partial v^2} \right)^2 \mathrm{d}v. \tag{3.7}$$

$\mathscr{P}$ is used to impose some level of smoothness to the LOSVD. The uncertainties of the resulting LOSVDs are estimated using Monte Carlo simulations. A galaxy spectrum is created by convolving the template spectrum with the measured LOSVD. Then 100 realizations of that initial galaxy spectrum are created by adding appropriate Gaussian noise. The LOSVDs of each realization are determined and used to specify the confidence intervals.

- The **Penalized Pixel Fitting** method (pPXF, Cappellari & Emsellem 2004) applies the general idea of MPL to the parametric expansion of LOSVDs in Gauss-Hermite polynomials. First the template and galaxy spectra are rebinned in $\ln \lambda$, then the best fit to the galaxy spectrum is found by convolving an optimal template spectrum with the corresponding LOSVD. As pPXF is very sensitive to template mismatch an optimal template spectrum is derived for each galaxy spectrum individually from a large library of templates spanning a broad range in age and metallicity. The best-fitting parameters (v, σ, h_3, ..., h_M) are determined by minimizing a penalized χ^2.

3.2 Spectral features in the K-band

To measure the kinematics of a galaxy both emission and absorption lines may be useful. Sharp emission lines originate from gas and are usually easy to detect, but gas is often shocked or non-gravitationally perturbed and might not follow the stellar potential. Therefore the gravitational potential and thus the black hole mass of the galaxy may not be reliably reconstructed using gas kinematics. Stars on the other hand can be treated as collisionless material and therefore their distribution reflects the gravitational potential much more reliably. Spectral absorption features often used to measure stellar kinematics are the Mg I b triplet ($\lambda\lambda 5164$, 5173, 5184 Å) or the strong near-infrared Ca II triplet (CaT, $\lambda\lambda 8490$, 8542, 8662 Å). Late-type galaxies and pseudobulges usually contain significant amounts of dust, therefore optical light from the nucleus is extincted and

Table 3.1: Most important spectral absorption features in the K-band

Species	λ (μm)
Na I	2.2062, 2.2090
Fe I1	2.2263
Fe I2	2.2387
Ca I	2.2614, 2.2631, 2.2657
Mg I	2.2814
^{12}CO(2 − 0)	2.2935
^{12}CO(3 − 1)	2.3227
^{12}CO(4 − 2)	2.3525
^{12}CO(5 − 3)	2.3829

the kinematics obtained at optical wavelengths might be unreliable. To minimize the extinction, galaxy spectra should be measured in a wavelength range as red as possible, but not redder than 3 μm where hot dust starts to dilute the stellar continuum. At 8500 Å, where the CaT is formed, the extinction A_{8500} is still ≈ 40% of that in the V-band, in addition the CaT lines are often blended with emission lines when the nucleus is active.

In the K-band the extinction effects are much less important ($A_{2.2\,\mu m} \approx 0.1\, A_V$). The strongest absorption features here are the CO bandheads redwards of 2.29 μm (see Table 3.1 for a list of the most important absorption features in the K-band) which show up when stars move into the red giant phase. The ^{12}CO(2-0) bandhead is by far the strongest absorption feature in the entire near-IR range sampled by SINFONI. Its blue edge is very sharp which makes it particularly sensitive to stellar motions. It is therefore well suited even for moderate S/N spectra to measure the dynamics of obscured galaxy nuclei. The bandheads are followed redwards by a number of other molecular lines belonging to the same transition, usually unresolved, which causes the particular asymmetric shape. Gaffney et al. (1995) showed that the CO(2 − 0) bandhead can be used to measure reliably the kinematics of galaxies if adequate template stars with the same CO equivalent width as the galaxy are chosen and the wavelength range around the bandhead is not too narrow.

The strength of the CO absorption feature depends on stellar parameters (Origlia & Oliva, 2000):

- Effective temperature T_{eff},
- Surface gravity g,
- Microturbulent velocity ξ,
- Metallicity and carbon abundance,

and therefore varies with spectral type and class. The depth of the bandheads increases with increasing metallicity and with decreasing T_{eff}. Strong CO features are present in regions of recent star formation, where they are produced by the population of very massive stars in the red supergiant phase $10^7 - 10^8$ years after the burst, and in old stellar populations (few Gyr) dominated by evolved late-type giants. The CO index, defined as the equivalent width of ^{12}CO(2-0) or the area between the normalised spectrum and the extrapolated continuum in the region of the bandhead, thus can be used to characterize a stellar population (Oliva et al., 1995). However, unambiguous interpretations in terms of e.g. age and metallicity is not yet possible due to the lack of a theoretical spectral synthesis model in the near-IR, equivalent to the Lick index system in the optical. Other near-IR line-indices (Na I, Fe I1, Fe I2, Ca I, Mg I see e.g. Förster Schreiber 2000; Frogel et al. 2001; Silva et al. 2008) can be defined similarly. They can be used to detect population differences for example due to nuclear discs, and thus help to understand the kinematical structure.

3.3 Stellar templates

A collection of stellar kinematic templates have been observed during the commissioning run of SINFONI in 2004, during the observation runs between March 2005 and September 2006 whenever the weather conditions were too bad for the normal program, or by other groups using SINFONI for similar purposes (6 templates, N. Neumayer, private communication). In total 18 template spectra are available, eleven of which for more than one platescale (see Tab. 3.2). Using templates observed with the same instrument as the galaxies has the advantage that the spectral resolution and the instrumental broadening is the same as for the galaxies. However, the number of templates is limited and in some cases it may not be possible to find matching templates for a galaxy. On the other hand, most libraries of stellar spectra in the K-band (e.g. Mármol-Queraltó et al. 2008 and references therein) have a lower spectral resolution than SINFONI or probe a limited wavelength range or range of spectral types. These are therefore not suitable, as for low-σ galaxies a sufficiently high spectral resolution is required.

Usually galaxies of an age $\gtrsim 1$ Gyr are dominated by evolved late-type giants, therefore mainly K or M type giants are suitable for the derivation of galaxy kinematics. Silge & Gebhardt (2003) showed that it is particularly important that the CO equivalent width of the templates matches that of the galaxy spectra. A non-matching equivalent width would induce a bias in σ. The CO equivalent widths of the templates and the galaxies were measured in the same way as described in Silge & Gebhardt (2003) by taking the area between 2.290 μm and 2.304 μm rest frame wavelength and dividing this by the continuum. The continuum is defined as a straight-line fit through two

Table 3.2: Stellar kinematic template stars.

Name	spectral type	M_K (mag)	CO (Å)	scale (mas)
HD012642	K5 III	1.753	12.7	25
HD013445	K1 V	4.125	3.0	25, 100, 250
HD049105	K0/1 III	5.206	8.0	25, 100, 250
HD075022	K2/3 III	4.300	12.0	25, 100
HD128598	K1 IV/V	5.177	9.0	25
HD132411	K3 V	7.437	4.5	100, 250
HD133507	K2 IIICNIb/II	4.082	8.8	100
HD141665	M5 III	1.449	15.8	100, 250
HD160346	K2 V	4.098	5.1	25
HD163755	K5/M0 III	0.750	15.6	25
HD179323	K2 III	2.642	11.1	25
HD181109	K5/M0 III	1.951	12.9	100, 250
HD198357	K3 III	2.272	10.8	25
HD201901	K3 III	2.001	11.4	100, 250
Hip54569	M0 V	6.972	5.7	100, 250
LHS513	M4 V	8.358	4.8	100, 250
SA112-0595	M0 III	7.296	12.8	100, 250
V2101 Oph	M5 II	5.846	16.7	100, 250

Note. The spectral types and K-band magnitudes are taken from Vizier (http://vizier.u-strasbg.fr/). In the last column the SINFONI platescales, in which the stars have been observed, are given.

points on both sides of the CO feature (median between 2.287 μm and 2.290 μm in the blue and the median between 2.304 μm and 2.307 μm in the red). As the continuum redwards of the CO feature is reduced by unresolved absorption lines of the same molecule, this approach is not very precise. For the measurement of small stellar population changes the continuum should be chosen more carefully. However, for the purpose of choosing the right templates this method is accurate enough, especially when programs are used that do not fit single templates but a combination of templates with different weights, like MPL (Gebhardt et al., 2000a). The CO equivalent widths of the templates are given in Tab. 3.2. They cover an equivalent width range between 3 and 17 Å. Fig. 3.2 shows the spectra of the stellar templates observed in the 100mas scale ordered by equivalent width. As demonstrated in Davies et al. (2007) and mentioned in Silge & Gebhardt (2003), most galaxies should have a CO equivalent width of around 12 Å with a range between 10 and 20 Å, and only when very young stellar populations (recent starburst) are present, CO is significantly lower.

Before trying to extract the LOSVDs, the CO equivalent widths of the galaxy spectra were measured in order to be able to select the most suitable templates (see Tab. 3.3 for the CO equivalent widths averaged over the SINFONI field of view of all galaxies discussed in this work). The velocity dispersion correction has been done according to Silge & Gebhardt (2003). For all galaxies discussed in Chapters 4-7 except NGC 5102 the equivalent width is in the expected range of $\sim 12 - 14$ Å. The equivalent width of NGC 5102 is significantly smaller and decreases from the outer parts with ~ 10 Å to the centre with ~ 5 Å. Thus a large fraction of the stars in this galaxy must be very young and have formed within the last few hundred Myrs. This is confirmed by the results of Davidge (2008) and references therein.

Knowing the CO equivalent width of galaxy and templates before extracting the kinematics is important, as it minimizes the risk of template mismatch in particular for methods, which only use one template star. Other methods are less sensitive to template mismatch, as they select a linear combination of several templates to obtain the best fit. Thus in cases where CO varies within the field of view these methods will automatically find the best matching template combination, while for the former methods one would have to choose manually the best template for each region in the galaxy. Nevertheless, even for methods that use template combinations it is useful to know the equivalent widths, so that templates can be excluded beforehand from the list.

3.4 Performance of FCQ and MPL in the near-IR

Most kinematic extraction techniques have been used mainly on optical absorption lines, but rarely in the near-IR. From all methods described above, those that can derive LOSVDs directly without

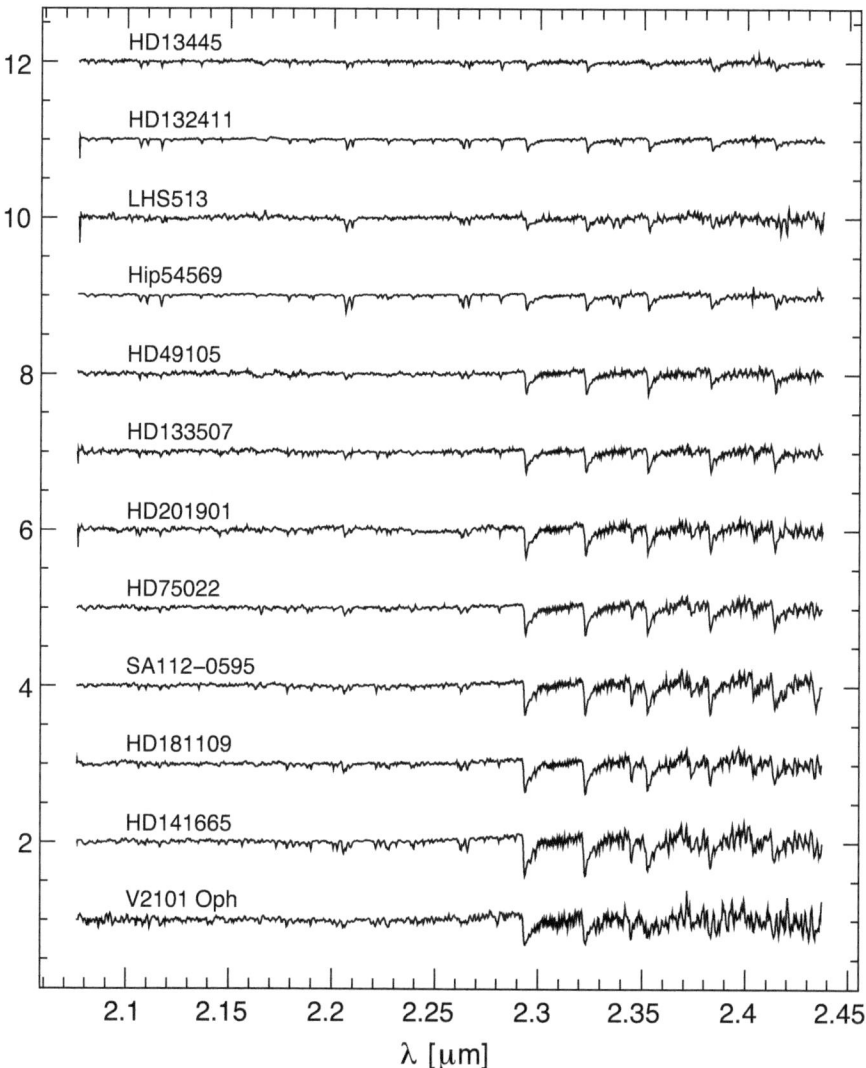

Figure 3.2: Stellar kinematic template stars for the 100mas scale, ordered by CO equivalent width (increasing from top to bottom). The spectra of the individual stars are continuum-normalised and shifted in vertical direction for a better comparison.

3.4. PERFORMANCE OF FCQ AND MPL IN THE NEAR-IR

Table 3.3: Mean CO equivalent width for all galaxies discussed in this work. The RMS errors are given in brackets.

Galaxy	CO (Å)
NGC 1316	13.8 (0.3)
NGC 3368	13.1 (0.3)
NGC 3489	14.3 (0.8)
NGC 4486a	12.4 (0.8)
NGC 5102	8.2 (1.1)

a priori assumption on the LOSVD shape (e.g. parametrization into Gauss-Hermite parameters v, σ, h_3, h_4) are most suitable for the purpose of dynamical modelling and reconstruction of the orbital structure, as parametrization might discard important features of the LOSVDs. Thus there are only few possible deconvolution algorithms left, of which the two most common ones are chosen: FCQ (Bender, 1990) and MPL (the version of Gebhardt et al. 2000a). Thus in the following both the FCQ and the MPL method are tested on model galaxy spectra using Monte Carlo simulations. This will give information about the suitability to extract the correct LOSVDs from near-IR spectra, in particular from the CO bandheads. Instead of comparing the resulting very large number of LOSVDs directly, Eq. 3.3 is fitted to the LOSVDs and then the moments v, σ, h_3 and h_4 are compared. Similar tests have been performed at smaller wavelengths by Joseph et al. (2001). They compared the FCQ algorithm and the MPL algorithm of Merritt (1997), which uses a different penalty function than the code of Gebhardt et al. (2000a), on *HST* STIS spectra in the CaT region. They found that for small galaxy velocity dispersions, h_4 measured with FCQ is significantly biased. FCQ seems to have difficulties recovering the true LOSVDs when the velocity dispersion of the galaxy is similar to the dispersion of the template (i.e. the instrumental resolution). As the instrumental resolution of SINFONI is quite high, this is probably not important for most of the observed galaxies, but this needs to be quantified by simulations. In addition FCQ and also MPL might show a different behaviour when they are used with the asymmetric CO bandheads.

3.4.1 FCQ

First a large number of model galaxy spectra were created by convolving a stellar template star spectrum (HD181109) with LOSVDs of different shapes and adding different amounts of noise. As an example, Fig. 3.3 shows the spectra of the original template and the template convolved with LOSVDs of different widths. 100 model galaxy spectra were created for eleven values of the S/N (between 10 and 200 per pixel), $\sim$ 25 different LOSVDs ($v = 0$ km s^{-1}, $\sigma = 30 - 300$ km s^{-1},

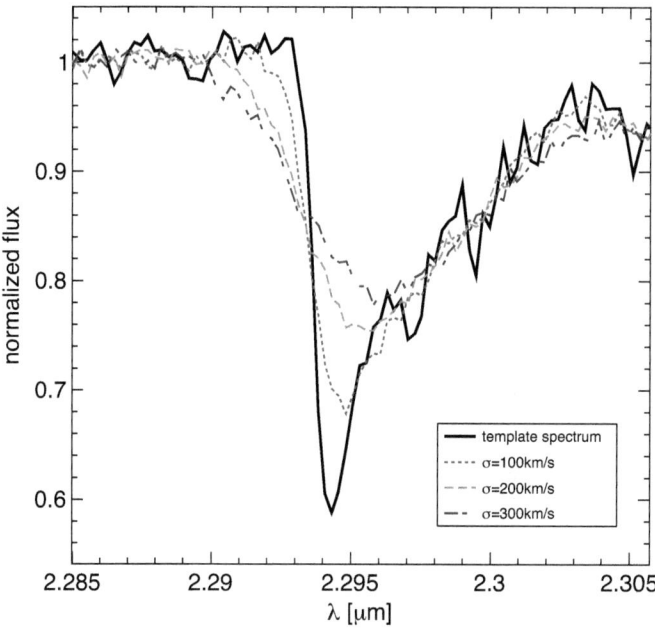

Figure 3.3: CO bandhead of the stellar kinematic template star HD181109 unconvolved (black dotted line) and convolved with Gaussian LOSVDs with velocity dispersions $\sigma = 100$, 200 and 300 km s^{-1} (coloured solid lines). The S/N of the convolved spectra is 140. The template spectrum is continuum-normalised.

$h_3 = h_4 = 0$) and four different wavelength ranges. Then the kinematics was extracted from these model galaxy spectra using FCQ with the original template star spectrum as a stellar template. For the Wiener Filter, which inhibits noise amplification during the deconvolution, the default "optimal" value $W = 1$ was chosen.

Fig. 3.4 shows the values (or, more precisely, the 68% confidence bands derived from the 100 model galaxy spectra created for each set of initial parameters) obtained with FCQ for v, $\Delta\sigma = \sigma_{out} - \sigma_{in}$, h_3 and h_4 as a function of S/N for five different velocity dispersions σ_{in} and four different wavelength ranges. The first thing to notice is the extremely strong bias in both $\Delta\sigma$ and h_4 when the K band CO bandheads are included. It is not possible to recover the correct dispersion for $\sigma_{in} < 100$ km s^{-1}: The smaller the dispersion, the stronger the bias, reaching ~ 60 km s^{-1} for $\sigma_{in} = 30$ km s^{-1}: The bias also strongly depends on the S/N. For the largest dispersion shown here, the bias becomes insignificant only at S/N~ 90. For dispersions larger than 150 km s^{-1} the

3.4. PERFORMANCE OF FCQ AND MPL IN THE NEAR-IR

minimum reliable S/N becomes slowly lower and is only ~ 35 for $\sigma_{in} = 250$ km s^{-1}. The bias in h_4 shows a different behaviour. The smaller the velocity dispersion and the S/N, the larger the bias in h_4. h_4 is always too small, with a bias between ~ -0.015 for $\sigma_{in} = 300$ km s^{-1} and ~ -0.15 for $\sigma_{in} = 30$ km s^{-1}. This is similar to the results of Joseph et al. (2001). The dependence on S/N is similar as for σ, but much less pronounced. For v and h_3 there is no bias and no dependence on σ or S/N. Just the error bars decrease with increasing S/N, which is observed for all parameters. It makes no difference whether all four K band CO bandheads are fitted or just the first two.

To find out if these strong biases depend on the spectral resolution the same simulations were done using a template spectrum observed with spectral dithering. The results are very similar. The biases in σ and h_4 depend in the same way on σ_{in}. The error bars are somewhat smaller and the $\sigma_{in} = 30$ km s^{-1} curve is less noisy. The dependence on S/N is less severe, as for $\sigma_{in} \gtrsim 150$ km s^{-1} reliable σ_{out} can be obtained already for S/N$\gtrsim 60$.

Omitting the CO bandheads and only fitting the spectral range between 2.0 and 2.29 µm reduces the biases in σ and h_4 significantly. The bias in σ only depends on the S/N, not on σ_{in} (except for extremely low S/N) any longer. The bias in h_4 is less strong (but still significant) and very weakly dependent on σ_{in} for $50 \leqslant h_4 \leqslant 300$ km s^{-1}. For smaller dispersions the bias is significantly stronger though not as strong as in the case with CO bandheads. On the other hand v and h_3 now seem to be biased. This bias disappears for high S/N$\gtrsim 90$ and depends only slightly on σ_{in} in the case of h_3.

As an alternative to the K-band CO bandheads it would as well be possible to use the H-band instead if it turned out that it is more suitable to obtain reliable kinematics. The strongest absorption features in the H-band are the second overtone CO absorption bands, which have a much more symmetric shape than the K-band CO bandheads, and the Si I 1.589 µm atomic absorption line. The right column in Fig. 3.4 shows the results of the simulations with H-band spectra. Unfortunately the biases do not disappear. The bias in σ depends much stronger on the S/N than for the K-band spectra. It does not, however, depend very strongly on σ_{in}. A significant bias due to σ only appears for very low $\sigma_{in} \lesssim 50$ km s^{-1}. The bias in h_4 is still present and depends in a similar way on S/N and σ_{in} than in the K-band.

The conclusion is that FCQ is not suitable for determining galaxy kinematics from the K-band CO bandheads. The majority of the observed galaxies analysed in this work has velocity dispersions < 150 km s^{-1} and S/N values between 40 and 100, exactly the range where FCQ does *not* work. The reason for the σ_{in} dependence seems to be the peculiar shape of the CO bandheads, as it vanishes when the CO bandheads are omitted from the fit. The bias in h_4 does not even disappear for very high σ_{in} and S/N, or when the CO bandheads are omitted. The amplitude is similar to

CHAPTER 3. STELLAR KINEMATICS

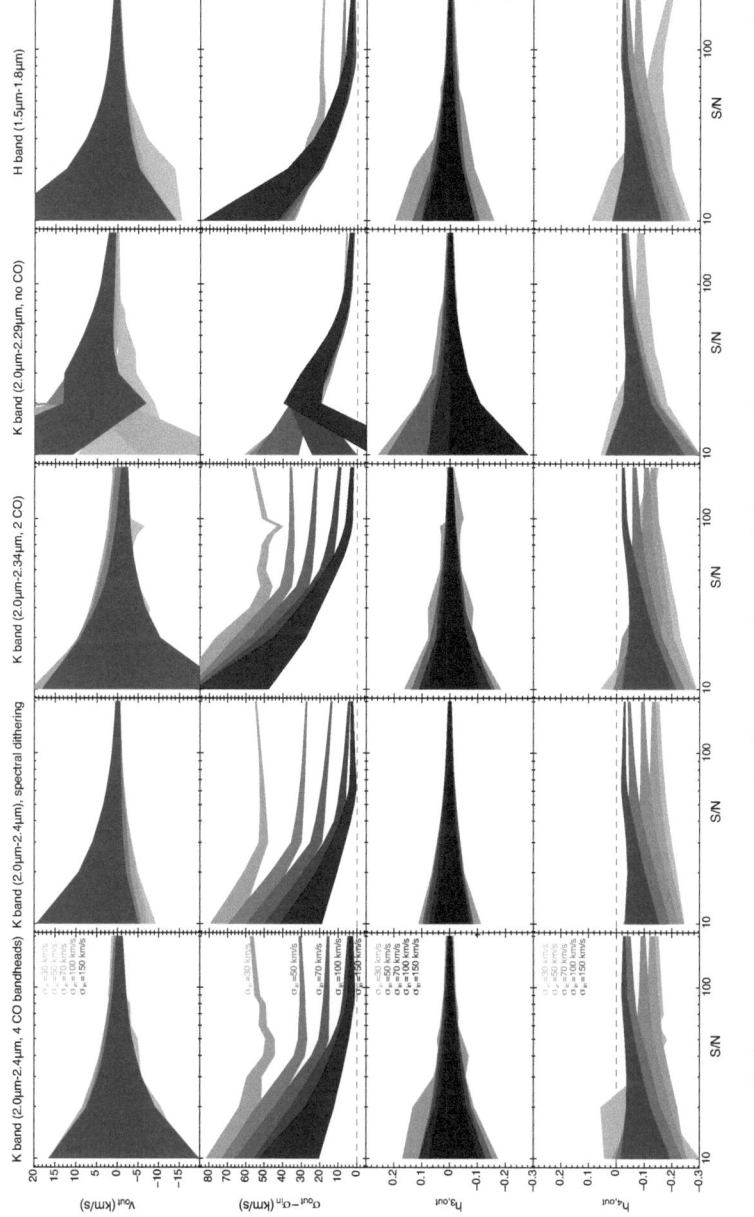

Figure 3.4: Stellar kinematic parameters, measured with FCQ, as a function of S/N and σ_{in} for a stellar template spectrum convolved with a Gaussian ($v_{in} = 0$, $\sigma_{in} = 30 - 150$ km s^{-1}, $h_{3,in} = 0$, $h_{4,in} = 0$) and different amounts of added noise (S/N = 10 − 200).

3.4. PERFORMANCE OF FCQ AND MPL IN THE NEAR-IR

the h_4 bias found in Joseph et al. (2001), who conclude that the dispersion of the galaxy needs to be significantly larger than the dispersion of the template when using FCQ.

3.4.2 MPL

The performance tests of MPL were done in a similar way as for FCQ. As the level of smoothing (i.e. the parameter α in Eq. 3.8) cannot be determined a priori, it is necessary also to test different values of α for each combination of LOSVD, S/N and wavelength range. This is particularly important in order to decide later which smoothing should be chosen for a certain data set. 100 model galaxy spectra were created for ~ 35 S/N-values between 1 and 170, many different LOSVDs ($v = 0$ km s^{-1}, $\sigma = 50 - 250$ km s^{-1}, $h_3 = -0.1; 0.0; 0.1$, $h_4 = -0.1; 0.0; 0.1$) and a few different wavelength ranges, and the kinematics of each spectrum was extracted using MPL with ≈ 100 different values for α between 0 and 500. The same stellar spectrum that was used to create the model galaxy spectra was used as kinematic template. The LOSVDs obtained with MPL by deconvolving galaxy and template spectra are binned with N bins in the velocity range from $-4.5\sigma_{in}$ to $+4.5\sigma_{in}$. The width of the LOSVD bins $\Delta v = 2 \times 4.5\sigma_{in}/N$ varies with σ_{in}. This assures that the FWHM region of each LOSVD is sampled by the same number of bins regardless how big or small the FWHM is. $N = 29$ was chosen for all tests, resulting in ~ 7.5 bins per FWHM of the LOSVD. Larger N do not improve the results. A problem is that the bin size becomes smaller than the instrumental resolution for small σ, which introduces biases, as shown below. The spectral range $\lambda = 2.275 - 2.349$ µm, i.e. the first two bandheads, is chosen for all tests except stated otherwise.

In the following the dependence of the measured LOSVDs on α, S/N, σ, h_3, h_4 and wavelength regions is tested. The subscript "in" (as in σ_{in}) is used to denote the initial value of a parameter, i.e. the value that was used for the convolution of the stellar spectrum with a LOSVD. The subscript "out" in turn denotes the measured value.

Dependence on S/N

First of all it was tested if there is a general dependence on S/N, as observed with FCQ. Fig. 3.5 shows the 68% confidence bands (derived from the 100 model galaxy spectra created for each set of initial parameters) obtained with MPL for v, $\Delta\sigma = \sigma_{out} - \sigma_{in}$, h_3 and h_4 as a function of the smoothing parameter α for different S/N values. The left and the middle column show the results for S/N= $10 - 140$ and Gaussian LOSVDs ($h_3 = h_4 = 0$) with $\sigma_{in} = 250$ km s^{-1}, the right column for S/N=$40 - 140$ and Gaussian LOSVDs with $\sigma_{in} = 80$ km s^{-1}. It seems that all parameters can be reliably reconstructed as long as the smoothing parameter is not too large. There is no obvious strong bias in any of the parameters as long as α is smaller than a few tens. The error bars increase with decreasing S/N, as expected. The only difference between $\sigma_{in} = 80$ km s^{-1} and $\sigma_{in} = 250$ km s^{-1} are the error bars, which are larger in h_3 and h_4 for $\sigma_{in} = 80$ km s^{-1}. A slight dependence on α can mainly be seen for σ_{out} and h_4. σ_{out} is a bit smaller than σ_{in} for small α, and increases somewhat with increasing α. It is generally consistent with σ_{in} within the errors. Only for very large values of the smoothing parameter $\alpha \gtrsim 40$, σ_{out} is not reliable anymore. h_4 depends stronger on α. For small α it is a bit too high, then continuously decreases slowly and at some point it starts to strongly decrease. Up to the point where the strong decrease starts, h_4 is still in agreement with 0 within the errors. This trend is seen for all S/N, but is most obvious for S/N $\geqslant$ 40 and also seems to be stronger for smaller σ. v and h_3 do not show any dependence at all on α or σ_{in}. They are constant with α and approximately 0 over a wide range.

Dependence on σ

In this paragraph the dependence of the obtained kinematics on the velocity dispersion σ_{in} was tested. The spectrum of HD181109 (100mas scale) was broadened with LOSVDs with the parameters $h_3 = h_4 = 0$ and $\sigma_{in} = 50 - 250$ km s^{-1} and many different noise values (S/N= 1 – 170). The results are shown in Fig. 3.6 for S/N= 140. The kinematic parameters for all σ_{in} values show the same trends with α and S/N as described above, but with certain differences between the different σ_{in} values. The error bars of h_3 and h_4 slightly decrease with increasing σ_{in}. σ_{out} decreases with increasing σ_{in}, such that for very small and very high σ_{in} it is difficult to find a value for α, where σ_{out} is not biased to higher (low σ_{in}) or smaller (high σ_{in}) values. For small σ_{in} the best smoothing parameter would be $\lesssim 1$, which is, unless the S/N is extraordinarily large, too small to get a smooth LOSVD. For high σ_{in} the best smoothing parameter would be $\sim 50 - 100$, which is already in the region where the measured kinematic parameters vary strongly and therefore are unreliable. This bias, however, is only of the order of $1 - 4$ km s^{-1}, which is usually within the errors for the S/N values of the observed galaxies.

A reverse trend is detected for h_4. For small $\sigma_{in} < 80$ km s^{-1} the measured h_4 is too small for all α values. For larger σ_{in} there is always an α range for which the measured h_4 is correct within the errors.

The reason for the dependence of σ_{out} and h_4 on σ_{in} is the ratio between the velocity resolution of SINFONI and the velocity bin width $\Delta v = 2 \times 4.5\sigma_{in}/N$ ($N = 29$). The spectral resolution of the three SINFONI platescales is $R = 4490$ (250mas), $R = 5090$ (100mas) and $R = 5950$ (25mas). This corresponds to FWHM velocity resolutions of 60 km s^{-1} (250mas), 53 km s^{-1} (100mas) and 45 km s^{-1} (25mas), or instrumental resolutions of 25 km s^{-1} (250mas), 23 km s^{-1} (100mas) and

3.4. PERFORMANCE OF FCQ AND MPL IN THE NEAR-IR

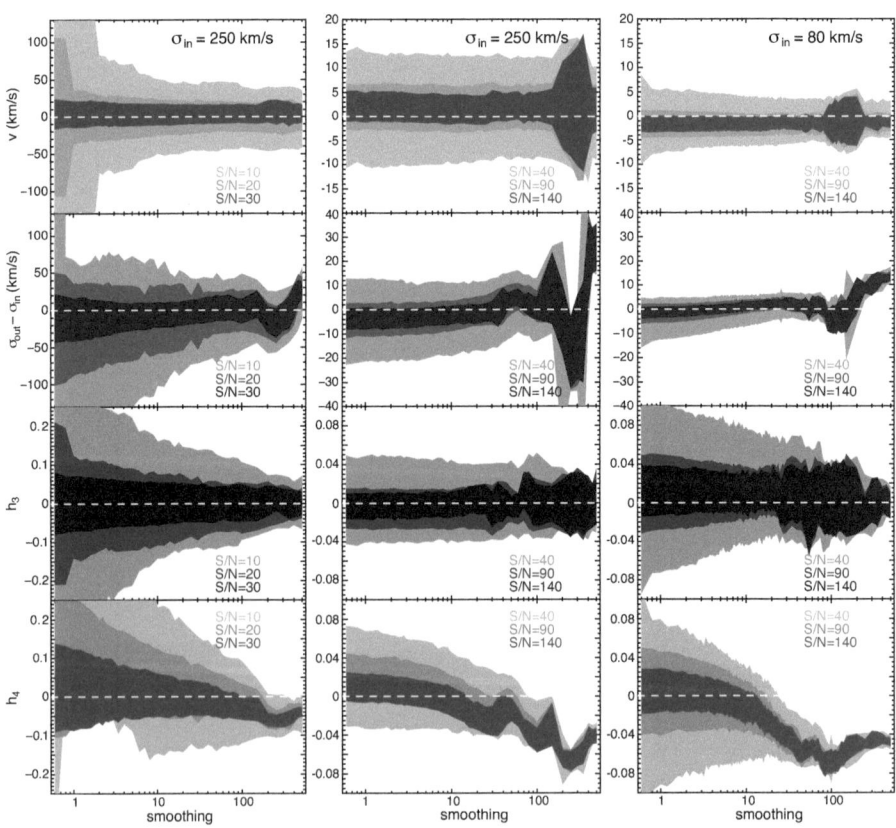

Figure 3.5: Kinematic parameters for different S/N values between 10 and 140. The kinematic template star HD12642 was used, observed on the 25mas scale. The spectrum was broadened using LOSVDs with $\sigma = 250$ (left and middle column) and 80 km s^{-1} (right column) and $h_3 = h_4 = 0$.

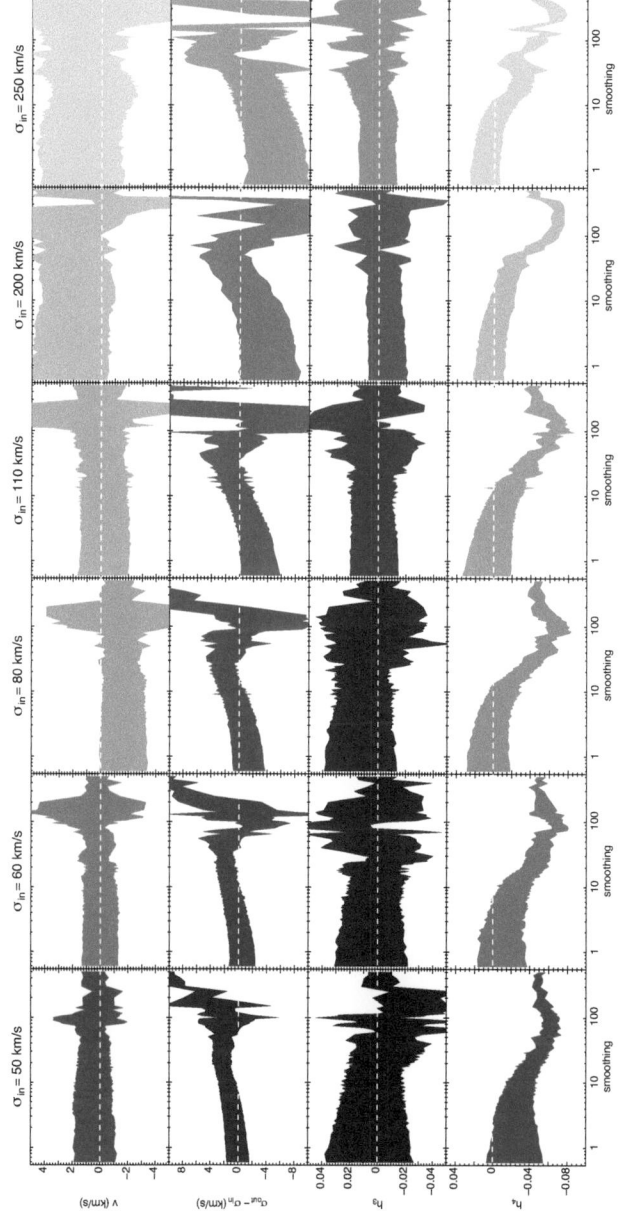

Figure 3.6: Kinematic parameters for different σ_{in} values. The star HD181109 observed on the 100mas scale was used, broadened by Gaussians with (from left to right) σ_{in} = 50, 60, 80, 110, 200 and 250 km s^{-1} and $h_3 = h_4 = 0$.

3.4. PERFORMANCE OF FCQ AND MPL IN THE NEAR-IR

19 km s^{-1} (25mas). For small dispersions Δv becomes smaller than the instrumental resolution. In fact, Δv is only ~ 16 km s^{-1} for $\sigma_{in} = 50$ km s^{-1}. If Δv is defined as above, the limiting σ_{in} down to which reliable kinematics can be obtained (i.e., where the velocity bin size equals the instrumental resolution) is 82 km s^{-1} for the 250mas scale, 73 km s^{-1} for the 100mas scale and 62 km s^{-1} for the 25mas scale. This is in agreement with the results from the simulations with 100mas scale spectra (Fig. 3.6), where the results are biased for $\sigma = 50, 60$ and 70 km s^{-1}, but not for $\geqslant 80$ km s^{-1}.

Thus it was tested if the measured kinematics for $\sigma_{in} = 50$ km s^{-1} changes when a larger binsize (realized by increasing the total velocity range $2 \times 4.5\sigma$ instead of changing N) is used (see Fig. 3.7). Doubling the binsize from 16 to 32 km s^{-1}, however, does not seem to produce the desired results. The LOSVDs are now sampled by half the number of bins as before, making it much more difficult to recover the exact LOSVD shape. σ can only be recovered for very small α close to 0, as σ_{out} increases strongly with α and has very small error bars. The bias in h_4 is even stronger for small α, also with much smaller error bars, but it is approximately constant (≈ -0.04) with α. v and h_3 are not affected except the error bars. Decreasing the binsize even further from 16 to 12 km s^{-1} has no effect on the results. Leaving the bin size as it was defined above, but using parametric fits instead of non-parametric fit can reproduce all parameters extremely well, but the errors in h_3 and h_4 are very large unless the S/N exceeds ~ 100 (Fig. 3.8). Therefore parametric fits could be used instead for very low-σ galaxies, if the S/N is high.

Dependence on h_3 and h_4

With the next set of simulations it was tested if the reconstructed kinematics depends on h_3 or h_4. This part is (slightly modified) taken from Appendix B of Nowak et al. (2008), where the case $\sigma = 250$ km s^{-1} is discussed. Model galaxy spectra using the star HD181109 (100mas) were created with $h_{3,in}$ and $h_{4,in}$ between -0.1 and $+0.1$ and $\sigma = 250$ km s^{-1}. The results are shown in Fig. 3.9 for S/N= 140. When $h_{3,in} \neq 0$, deviations from the "normal" (i.e. when $h_{3,in} = h_{4,in} = 0$) behaviour are present with increasing α mainly in v and $h_{3,out}$. For positive $h_{3,in}$ the velocity increases with increasing α and for negative $h_{3,in}$ the velocity decreases. $h_{3,out}$ is constant for a wide range of α values. Only for large α, for which all parameters show more or less strong deviations, $h_{3,out}$ deviates much stronger from the correct value than in the case of $h_{3,in} = 0$. When $h_{4,in} \neq 0$, deviations from the "normal" behaviour emerge with increasing α in σ_{out} and $h_{4,out}$. The larger $h_{4,in}$, the stronger the increase in σ_{out} and the decrease in $h_{4,out}$. In addition, with decreasing $h_{4,in}$ a bias to too positive values in $h_{4,out}$ is introduced and the slope of the $h_{4,out}$-α curve flattens.

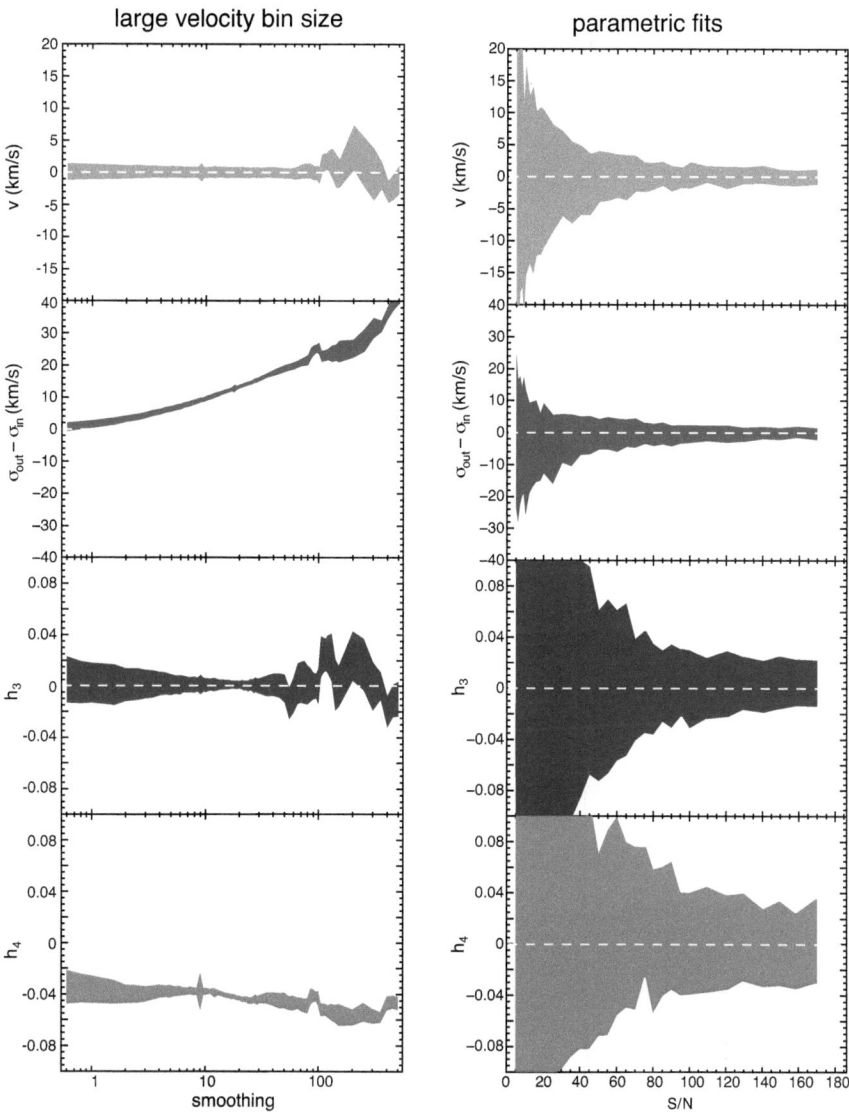

Figure 3.7: Kinematic parameters for $\sigma_{in} = 50$ km s^{-1} with a doubled velocity bin size $\Delta v = 4 \times 4.5\sigma_{in}/29$.

Figure 3.8: Kinematic parameters for $\sigma_{in} = 50$ km s^{-1} derived with a parametric fit instead of a non-parametric fit.

3.4. PERFORMANCE OF FCQ AND MPL IN THE NEAR-IR

h_3 and h_4 depend only slightly on each other. If $h_{3,\text{in}}$ and $h_{4,\text{in}}$ are $\neq 0$ *and* $h_{4,\text{in}} < 0$, the slope of the $h_{3,\text{out}}$-α curves are somewhat steeper and the bias in $h_{4,\text{out}}$ is somewhat stronger than in the $h_{3,\text{in}} = 0$ case.

Dependence on the wavelength region

There do not seem to be major differences between the kinematics derived from different wavelength regions. Different tests were made, using (1) the entire K-band (i.e., the region between 2.08 and 2.4 µm) and (2) only the first CO bandhead. The behaviour of v, σ, h_3 and h_4 (Fig. 3.10) shows no systematic changes to the case where the first two bandheads are used. Only the error bars change. Fitting the entire K-band spectrum seems to be the better choice for the derivation of reliable kinematics, as the errors are very small. However, the simulations are done with a single star using itself as a perfectly matching template. Real galaxies are much more complicated. The absorption lines bluewards of the CO bandheads (e.g. Na I, Ca I, Fe I, Mg I) can be blended with absorption lines of other elements. This is often the case for the Na I absorption line, which in stars usually has a small equivalent width, but in galaxies its equivalent width is much larger due to a contribution of silicon (Silva et al., 2008). The third and the fourth bandhead are often strongly affected by residual sky lines, which also can alter the measured kinematics. To illustrate this, Fig. 3.11 shows the fit of a template spectrum to a model galaxy spectrum generated using this template, and for comparison the fit of the best combination of templates found by MPL to a real galaxy spectrum. Many of the absorption lines cannot be fitted very well and the third and fourth bandhead are obviously strongly affected by residual sky lines. Thus in order to obtain reliable LOSVDs it is certainly better to accept larger error bars and just fit the first two bandheads.

Results

Taking all simulations together it can be confidently concluded that with MPL reliable kinematics can be obtained from the first two K-band CO bandheads of the SINFONI data when α is evaluated carefully for each data set. Due to the consistency of the results for different wavelength regions and the above mentioned problems for absorption features other than the first two CO bandheads, all LOSVDs of real galaxy spectra will be extracted from the first two bandheads only.

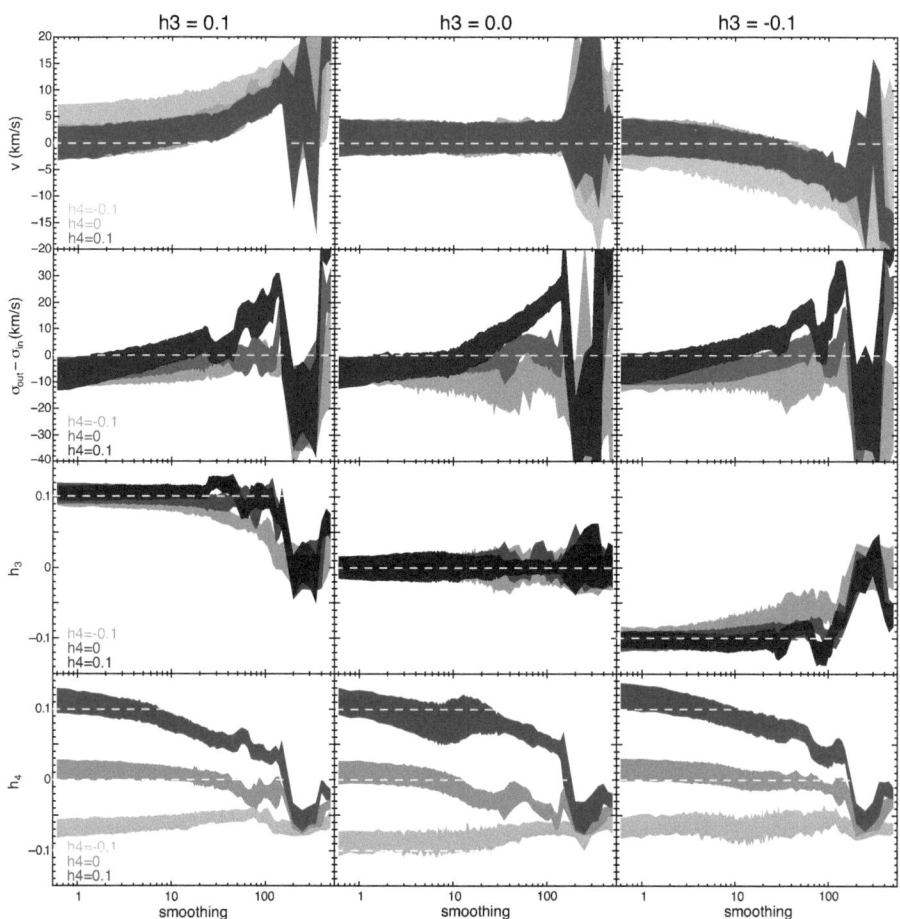

Figure 3.9: Kinematic parameters for different h_3 and h_4. Here the star HD181109 observed on the 100mas scale was used. It was broadened by LOSVDs with $\sigma = 250$ km s^{-1}, $h_3 = 0.1$, 0 or -0.1, $h_4 = 0.1$, 0 or -0.1 and S/N= 140.

3.4. PERFORMANCE OF FCQ AND MPL IN THE NEAR-IR

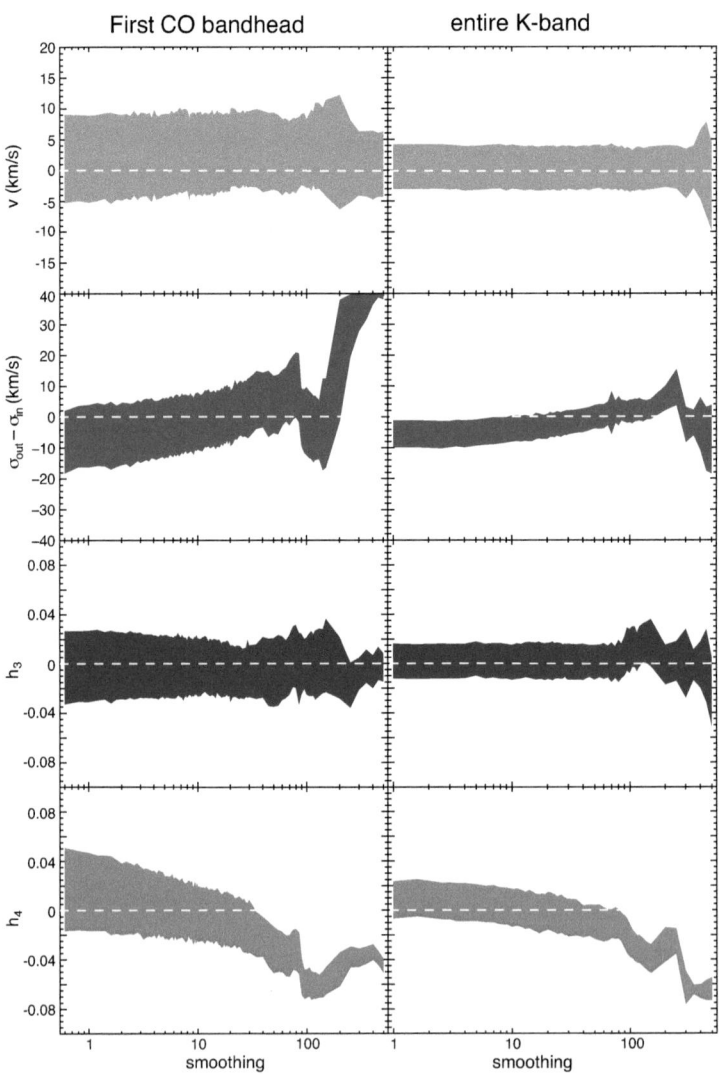

Figure 3.10: Kinematic parameters for different wavelength regions (left column: first CO bandhead only; right column: entire K-band). The spectrum of HD181109 was broadened by a Gaussian with $\sigma = 250$ km s^{-1}.

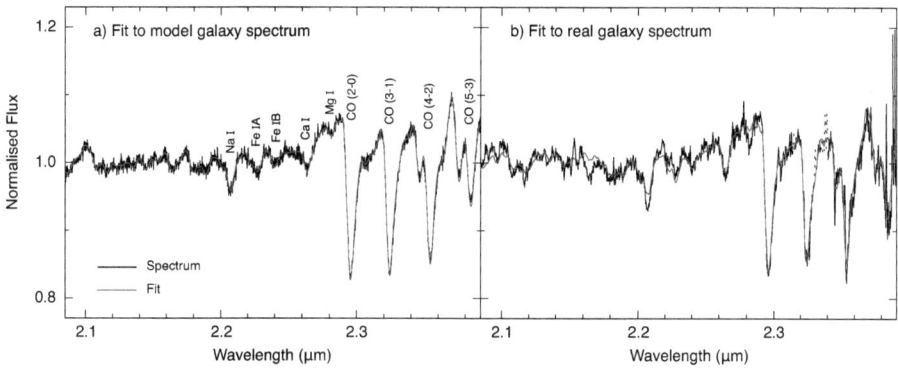

Figure 3.11: Panel (a) shows the fit to a model galaxy spectrum generated from the spectrum of the star HD181109, using the unconvolved spectrum of HD181109 as a template. Panel (b) shows the fit to a real galaxy spectrum, using the best combination of available 100mas stars.

3.5 Recipe to obtain LOSVDs from K-band SINFONI spectra

Prior to the usage of MPL all spectra (the galaxy spectra and the template spectra) have to be continuum-normalised. For that purpose a higher-order polynomial is fitted to the spectrum and the spectrum is then divided by the fit. The continuum normalisation should be done in the same way for galaxy and template spectra. Strong emission or absorption features are excluded from the fit.

Obtaining the stellar kinematics of a galaxy with MPL is an iterative process. The main parameters needed to determine the appropriate smoothing, σ and S/N (and to some extent also h_3 and h_4 if they are significantly different from zero), are not known a priori. Thus the first step is to determine the S/N. This is also necessary if the data should be binned in a way that the resulting bins have approximately all the same S/N (Voronoi binning, Cappellari & Copin 2003). MPL measures the S/N automatically as 1/rms of the fit to the spectrum. This fit is usually good as long as the template CO equivalent width matches the galaxy CO equivalent width (this is usually the case as MPL automatically chooses the best template combination), almost independent of α. Values for v and σ can be found in the literature for most galaxies, which are usually a good first guess to determine the initial velocity profile. If the kinematics is extracted from unbinned data, the S/N per pixel is low (between a few tens in the centre and a few in the outer parts, with S/N< 30 for most of the field of view), thus to determine a first good estimate of v and σ without too much scatter a high $\alpha \sim 50$ should be chosen (see Fig. 3.5).

3.5. RECIPE TO OBTAIN LOSVDS FROM K-BAND SINFONI SPECTRA

Once having the S/N information the spectra can be binned into higher S/N bins, either using Voronoi binning or the radial and angular binning scheme used for the dynamical models described in Chapter 4. The size of the spatial bins should be a compromise between a S/N which is as high as possible and good enough spatial resolution to resolve the sphere of influence. This usually means that the S/N of the central spaxels is taken as a reference such that the central bins comprise only one spaxel and the size of the other bins increases with radius. If the central S/N is very low, the size of the central bins might be increased, but should not exceed the size of the PSF or (as a strict upper limit) the size of the sphere of influence. From Fig. 3.5 it becomes clear that, in order to get relatively noise-free LOSVDs with reasonably small error bars, the S/N should be at least ~ 40. The luminosity-weighted spatial binning of the spectra was done using the IRAF[1] task imcombine.

For the binned spectra the S/N then should also be exactly determined for each bin. This is not so much important for the determination of the best smoothing parameter, for which an approximate value would be sufficient. The exact S/N is needed for the determination of the LOSVD error bars, which is done by creating 100 spectra from the templates convolved with the measured LOSVDs and appropriate amounts of noise added to them. Adding too much or too few noise would falsify the error bars. The average velocity and velocity dispersion determined from the unbinned data can be used to define the velocity range and bin width. $\alpha = 10$ is a reasonable upper limit for the high-S/N, binned spectra for this first iteration, based on the simulations in the previous section. In case there are residual sky lines, bad pixels, or cosmics in the region between or blue or redwards of the CO bandheads the estimated S/N is too low. In these cases it is advisable to restrict the wavelength range or to use only one bandhead.

Having a good estimate for the S/N and σ, it is now the time to analyse the simulations to determine the most appropriate value of α for each bin of each data set. The measured σ and h_4 are the only parameters that depend on α, as long as the galaxy h_3 and h_4 do not differ too much from 0. Thus the best value for α can be chosen as where $\sigma_{out} - \sigma_{in} = 0$ or $h_{4,out} - h_{4,in} = 0$. The problem is, that these two values can differ substantially. Fig. 3.12 shows for high and low σ the "best" smoothing parameters as a function of S/N determined from σ_{out} and from $h_{4,out}$. This "best" α has been determined as the minimum χ^2 difference between the measured and the correct σ or h_4 for $\alpha < 40$. The scatter is quite large, but the overall trend with S/N can be described by a single or double exponential. α_σ is always larger than α_{h_4}. Both values decrease with decreasing σ_{in} and increasing S/N of the galaxy. However, the bias in σ_{out} is small and within the errors of the real

---

[1] IRAF (http://iraf.noao.edu/) is distributed by the National Optical Astronomy Observatories, which are operated by the Association of Universities for Research in Astronomy, Inc., under cooperative agreement with the National Science foundation.

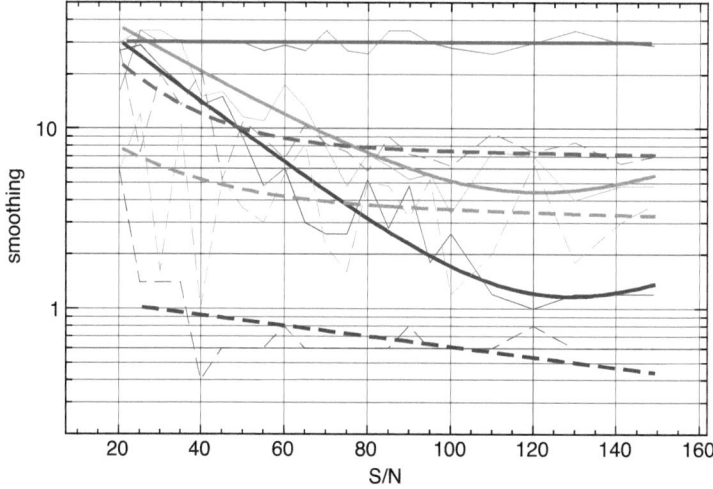

Figure 3.12: "Best" smoothing parameter α as a function of S/N for $\sigma = 250$ km s^{-1} (red), 80 km s^{-1} (green) and 50 km s^{-1} (blue). α derived from the χ^2 difference between measured and correct σ are plotted as thin solid lines, single or double exponential fits to these data points are plotted as thick solid lines. α derived in the same way from h_4 and the according fits are plotted as dashed thin and thick lines.

3.5. RECIPE TO OBTAIN LOSVDS FROM K-BAND SINFONI SPECTRA

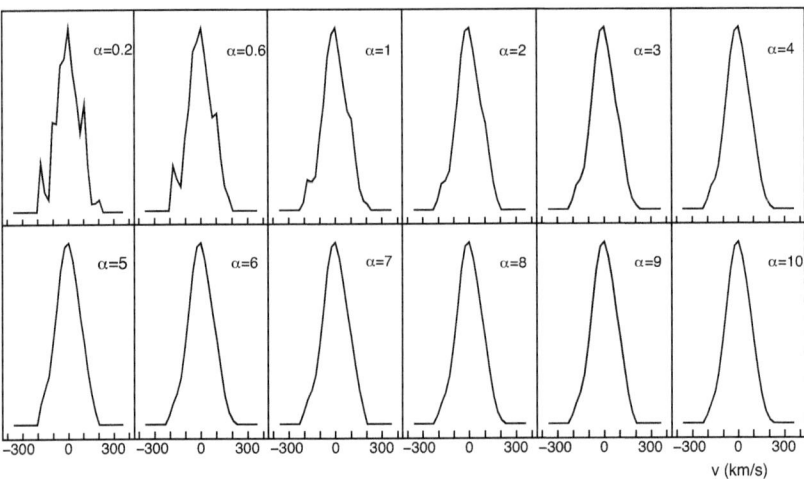

Figure 3.13: LOSVDs ($\sigma = 80$ km s^{-1}) for $\alpha = 0.2 - 10$ at a S/N of 70. The noise in the LOSVDs decreases with increasing α.

value when the chosen α is smaller than α_σ for most S/N values of our SINFONI observations. If, on the contrary, the chosen α is larger than α_{h_4}, the bias in h_4 could be large and not within the errors of the real h_4, as h_4 at some point starts to decrease very strongly. Therefore the final α should be closer to α_{h_4} than to α_σ. In addition to the Gauss-Hermite parameters it is also important to look at the LOSVDs. The purpose of the smoothing is to reduce the noise as much as possible, thus blindly choosing the best α indicated by h_4 does not help if it is still so small that the LOSVDs are noisy. Fig. 3.13 shows some example LOSVDs ($\sigma = 80$ km s^{-1}) for different α at S/N= 70. For $\alpha \leqslant 1$ they are very noisy, but for larger α the noise is insignificant, so α_{h_4} seems to be a good choice. Note that real data may have different noise patterns (e.g. due to sky residuals, bad pixels etc.), thus it might be necessary to choose $\alpha > \alpha_{h_4}$. For $\alpha_{h_4} \leqslant \alpha \lesssim 10$ the measured h_4 is usually in agreement with the correct h_4 within the errors, as long as σ is not too small.

For velocity dispersions smaller than 80 km s^{-1} the situation is much more difficult, because h_4 is biased for all α to smaller values. The reason is the velocity bin width, which is smaller than the instrumental resolution, when using a non-parametric fit. As changing parameters such as the velocity bin width does not help to reduce the bias, one possible solution is to (1) derive the h_4 bias for the velocity dispersion range of the galaxy from simulations, (2) subtract the bias from h_4 in the simulations and determine an appropriate α from the bias-corrected h_4 values, (3) determine the LOSVDs of the galaxy spectra with this α, (4) do a Gauss-Hermite fit to the LOSVDs, (5) subtract the previously determined bias from the measured h_4 values and (6) calculate corrected LOSVDs from the corrected Gauss-Hermite parameters. If the last three steps are omitted and the uncorrected LOSVDs of the galaxy are modelled, the galaxy models would be tangentially biased, which in turn could result in a biased SMBH mass. Gauss-Hermite shaped LOSVDs on the other hand, determined in the described way, may not represent the real velocity profiles anymore, thus the mass of the black hole might be biased in an unforseen way. In case the S/N is very high, a parametric fit would possibly be a more reliable alternative.

Having now fixed all necessary parameters, the LOSVDs of the galaxy spectra can finally be determined. For small S/N or a large S/N range an α that varies with S/N might be chosen, otherwise a constant α will be sufficient. As for the kinematic template stars, the MPL code automatically selects those templates that best match the galaxy spectrum. Nevertheless, in order to avoid a possible source of errors, it might be useful to a priori exclude those templates that have a CO equivalent width that differs from the galaxy CO equivalent width by several Angstroms. It might also be useful to extract the LOSVDs first with only one template and to check the fits and the LOSVDs. In this way all unsuitable templates can be excluded and variations of the kinematics with template or template combination can be assessed.

The final step is the determination of the 68% errors of the LOSVDs. For this purpose the stellar template spectrum is convolved with the previously measured LOSVD. Of this initial galaxy spectrum 100 versions are created by adding an appropriate amount of noise. The determination of the S/N value needed for this step has been described above. For each realization of the galaxy spectrum the LOSVD is measured and from this distribution of LOSVDs the 68% confidence bands are determined.

4

The supermassive black hole in NGC 4486a detected with SINFONI at the VLT[1]

Abstract

The near-infrared integral-field spectrograph SINFONI at the ESO VLT opens a new window for the study of central supermassive black holes. With a near-IR spatial resolution similar to *HST* optical and the ability to penetrate dust it provides the possibility to explore the low-mass end of the $M_\bullet$-σ relation ($\sigma < 120$ km s^{-1}) where so far very few black hole masses were measured with stellar dynamics. With SINFONI we observed the central region of the low-luminosity elliptical galaxy NGC 4486a at a spatial resolution of $\approx 0.1''$ in the K band. The stellar kinematics was measured with a maximum penalised likelihood method considering the region around the CO absorption bandheads. We determined a black hole mass of $M_\bullet = (1.25^{+0.75}_{-0.79}) \times 10^7$ M$_\odot$ (90 % confidence limit) using the Schwarzschild orbit superposition method including the full 2-dimensional spatial information. This mass agrees with the predictions of the $M_\bullet$-σ relation, strengthening its validity at the lower σ end.

---

[1]This chapter has been published as "Nowak N., Saglia R. P., Thomas J., Bender R., Pannella M., Gebhardt K., Davies R. I., 2007, MNRAS, 379, 909"

4.1 Introduction

Studies of the dynamics of stars and gas in the nuclei of nearby galaxies during the last few years have established that all galaxies with a massive bulge component contain a central supermassive black hole (SMBH; Bender & Kormendy 2003; Richstone et al. 1998). Masses of these SMBHs ($M_\bullet$) are well correlated with the bulge luminosity or mass, respectively, and with the bulge velocity dispersion σ (Ferrarese & Merritt, 2000; Gebhardt et al., 2000b; Kormendy & Richstone, 1995). There are, however, still a lot of open questions in conjunction with the $M_\bullet$-σ correlation, among them the exact slope, its universality and the underlying physics. A more precise knowledge of the behaviour of the $M_\bullet$-σ relation would help to constrain theoretical models of bulge formation and black hole growth (e.g. Burkert & Silk 2001; Haehnelt & Kauffmann 2000; Silk & Rees 1998 and others).

In inactive galaxies, the evidence for the existence of black holes and their masses comes from gravitational effects on the dynamics of stars inside the black hole's sphere of influence. Since the radius of the sphere of influence scales with the black hole mass $M_\bullet$, high resolution observations are needed to detect SMBHs in the low-mass regime, even for the most nearby galaxies. Further difficulties arise from the presence of dust in many of these galaxies, particularly in discs, which enforces observations in the infrared. Up to now the low-mass regime ($\sigma \lesssim 120$ km s^{-1}) is sparsely sampled with only four stellar dynamical black hole mass measurements (Milky Way, e.g. Gillessen et al. 2009; M32, Verolme et al. 2002; NGC 4258 Siopis et al. 2009; NGC 7457 Gebhardt et al. 2003) and some upper limits (e.g. Gebhardt et al. 2001; Valluri et al. 2005). Since the near-infrared integral-field spectrograph SINFONI became operational (Bonnet et al., 2004; Eisenhauer et al., 2003a), it is now possible to detect low-mass black holes in dust-obscured galaxies at a spatial resolution close to that of *HST*.

NGC 4486a is a low-luminosity elliptical galaxy in the Virgo cluster at a distance of 16 Mpc, close to M87. It contains an almost edge-on nuclear disc of stars and dust (Kormendy et al., 2005). The bright star $\sim 2.5''$ away from the centre makes it impossible to obtain undisturbed spectra with conventional ground-based longslit spectroscopy. However, it is one of the extremely rare cases, where an inactive, low-luminosity galaxy can be observed at diffraction limited resolution using adaptive optics with a natural guide star (NGS). This feature made NGC 4486a one of the most attractive targets during the years between the commissioning of SINFONI and similar instruments and the installation of laser guide stars (LGS). NGC 4486a is the first of our sample of galaxies observed or planned to be observed using near-infrared integral-field spectroscopy with the goal to tighten the slope of the $M_\bullet$-σ relation in the low-σ regime ($\lesssim 120$ km s^{-1}) and for pseudobulge galaxies.

This chapter is structured as follows: In Section 4.2 we present the observations and the data reduction. The derivation of the kinematics and the photometry is described in Sections 4.3 and 4.4 and the dynamical modelling procedure is explained in Section 4.5. In Section 4.6 the results are presented and discussed and conclusions are drawn.

4.2 Observations and data reduction[2]

NGC 4486a was observed on April 5 and 6, 2005, as part of guaranteed time observations with SINFONI (Bonnet et al., 2004; Eisenhauer et al., 2003a) at the Very Large Telescope (VLT) UT4. SINFONI consists of the integral-field spectrograph for IR faint field imaging (SPIFFI) (Eisenhauer et al., 2003b) and the Multi-Application Curvature Adaptive Optics (MACAO) module (Bonnet et al., 2003). We used the K band grating (1.95 − 2.45 μm) and the 3″ × 3″ field of view (0.05″ × 0.1″ px^{-1}). The bright ($R \approx 11$ mag) star located $\sim 2.5″$ southwest of the nucleus was used for the AO correction. The seeing indicated by the optical seeing monitor was between 0.6″ and 0.85″, resulting in a near-infrared seeing better than $\sim 0.7″$, which could be improved by the AO module to reach a resolution of 0.1″ ($\sim 25\%$ Strehl). For the chosen configuration SINFONI delivers a nominal FWHM spectral resolution of $R \approx 5090$. In total 14 on-source and 7 sky exposures of 600 s each were taken in series of "object-sky-object" cycles, dithered by up to 0.2″. During the last observation block the optical seeing suddenly increased above 1.0″. Therefore, to keep the spatial resolution as high as possible, only exposures with a seeing $\lesssim 0.85″$ were considered for the derivation of the kinematics. In order to derive the point-spread function (PSF) an exposure of the AO star was taken regularly.

The SINFONI data reduction package spred (Abuter et al., 2006; Schreiber et al., 2004) was used to reduce the data. It includes all common reduction steps necessary for near-infrared data plus routines to reconstruct the three-dimensional data cubes. After subtracting the sky frames from the object frames, the data were flatfielded, corrected for bad pixels, for distortion and wavelength calibrated using a Ne/Ar lamp frame. The wavelength calibration was corrected using night-sky lines if necessary. Then the three-dimensional data cubes were reconstructed and corrected for atmospheric absorption using the B9.5V star Hip 059503. As a final step all data cubes were

---

[2]Based on observations at the European Southern Observatory (ESO) Very Large Telescope (VLT) (075.B-0236).

4.2. OBSERVATIONS AND DATA REDUCTION

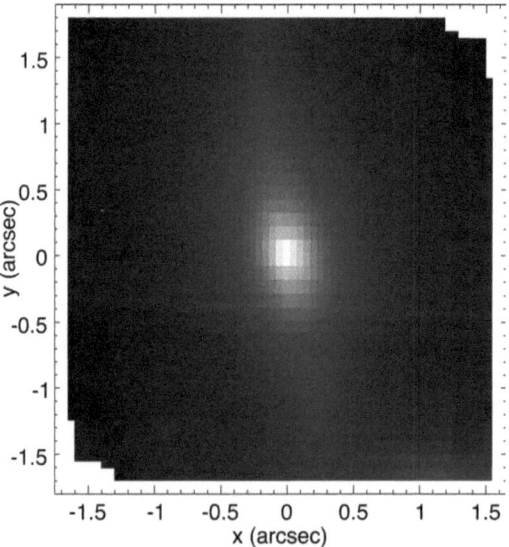

Figure 4.1: SINFONI image of NGC 4486a.

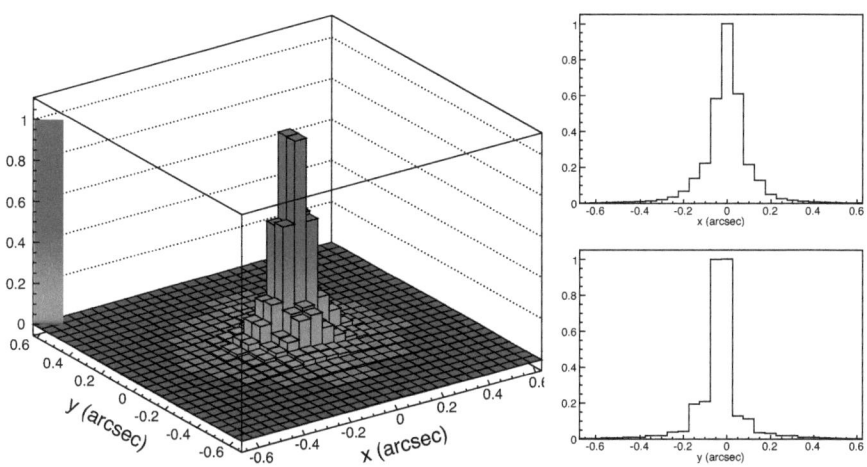

Figure 4.2: Left: The two-dimensional PSF. Right: x- and y-profiles of the PSF.

averaged together to produce the final data cube. Fig. 4.1 shows the resulting SINFONI image (collapsed cube) of NGC 4486a. The data of the telluric and the PSF stars were reduced likewise.

As PSF we take the combined and normalised image of the PSF star exposures (see Fig. 4.2). Its core can be reasonably well fitted by a Gaussian profile with a FWHM of $\sim$ 100 mas (7.7 pc). The diameter of the sphere of influence of the black hole can be roughly estimated to 130 mas using the $M_\bullet$-σ relation of Tremaine et al. (2002) and is therefore resolved.

4.3 Kinematics

The kinematic information was extracted using the maximum penalised likelihood (MPL) technique of Gebhardt et al. (2000a), which obtains non-parametric line-of-sight velocity distributions (LOSVDs). As kinematic template stars we use six K0 to M0 stars which were observed during commissioning and our guaranteed time observations in 2005 with SINFONI using the same configuration as for NGC 4486a. Both galaxy and template spectra were continuum-normalised. An initial binned velocity profile is convolved with a linear combination of the template spectra and the residuals of the resulting spectrum to the observed galaxy spectrum are calculated. The velocity profile is then changed successively and the weights of the templates are adjusted in order to

4.3. KINEMATICS

optimise the fit to the observed spectrum by minimizing the function $\chi_p^2 = \chi^2 + \alpha \mathcal{P}$, where α is the smoothing parameter that determines the level of regularisation, and the penalty function $\mathcal{P}$ is the integral of the square of the second derivative of the LOSVD. We fitted only the first two bandheads CO(2−0) and CO(3−1). The higher-order bandheads are strongly disturbed by residual atmospheric features. At wavelengths $\lambda < 2.29$ μm the absorption lines are weak and cannot be fitted very well by the templates.

The uncertainties on the velocity profiles were estimated using Monte Carlo simulations (Gebhardt et al., 2000a). A galaxy spectrum is created by convolving the template spectrum with the measured LOSVD. Then 100 realizations of that initial galaxy spectrum are created by adding appropriate Gaussian noise. The LOSVDs of each realization are determined and used to specify the confidence intervals.

In order to test the performance of the method on our SINFONI data and to find the best fitting parameters we performed Monte Carlo simulations on a large set of model galaxy spectra (see Section 3.4.2). These were created from stellar template spectra by convolving them with both Gaussian and non-Gaussian LOSVDs and by adding different amounts of noise. We found that the reconstructed LOSVDs resemble the input LOSVDs very well if the smoothing parameter α is chosen adequately (Joseph et al., 2001; Merritt, 1997). The best choice of α depends on the signal-to-noise ratio (S/N) and the velocity dispersion σ of the data. To maximize the S/N of the data a binning scheme with 11 radial and 5 angular bins per quadrant, similar to that used in Gebhardt et al. (2003), was chosen. The centres of the angular bins are at latitudes $\vartheta = 5.8°, 17.6°, 30.2°, 45.0°$ and $71.6°$ from the major to the minor axis. The bins are not overlapping, but spatial resolution elements at the border between bins may be divided into parts where each part is counted to a different bin. The spectra within each bin were averaged with weights according to their share in the bin. The radial binning scheme ensures that an adequate S/N level comparable to that of the central spectrum (S/N $\approx$ 40) is maintained at all radii at the cost of spatial resolution outside the central region.

The resulting two-dimensional kinematics (v, σ, h_3, h_4) is presented in Fig. 4.3. It illustrates the superposition of the kinematics of the two distinct components in NGC 4486a – the disc and the bulge. Whereas the velocity map shows a regular rotation pattern, the cold stellar disc can be clearly distinguished from the surrounding hotter bulge in the velocity dispersion map. The velocity dispersion of the disc is $\approx 20 - 30$ km s^{-1} smaller than that of the bulge. Asymmetric and symmetric deviations from a Gaussian velocity profile are quantified by the higher-order Gauss-Hermite coefficients h_3 and h_4 (Gerhard, 1993; van der Marel & Franx, 1993). Fig. 4.4 shows the kinematic profiles of NGC 4486a along the major axis at angles $\theta = +5.8°$ and $\theta = -5.8°$.

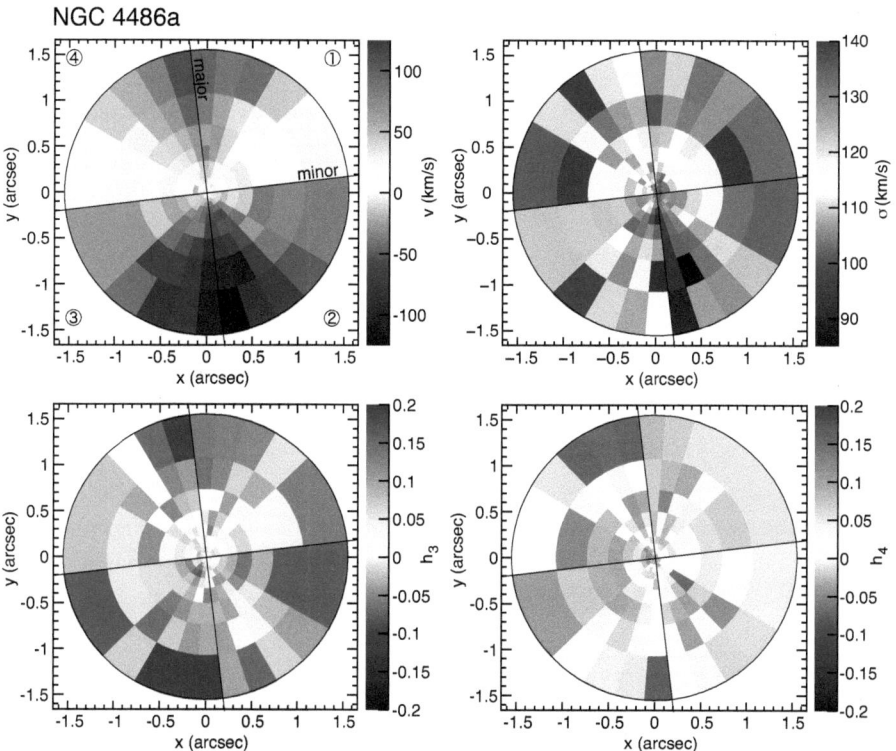

Figure 4.3: Two-dimensional stellar kinematics (v, σ, h_3, h_4) of NGC 4486a. Major axis, minor axis and the numbering of the quadrants are indicated in the velocity map (upper left).

4.4. IMAGING

The $+5.8°$-profile agrees very well with the adjacent $-5.8°$-profile within the error bars. When comparing the profiles at negative radii with the profiles at positive radii slight asymmetries can be seen (especially in σ), but as the errors are relatively large, deviations from axisymmetry are small.

The star at $2.5''$ from the centre, which heavily dilutes optical spectra of NGC 4486a taken without AO, does not have any significant effect on the kinematics derived here. Quadrant 2 is affected most, as it is located in the direction of the star. The fraction of inshining light from the star is only about 13% at $0.7''$ from the centre of the galaxy in the direction of the star (the outermost point covered by the exposures of the PSF star) due to the narrow PSF. In addition the spectrum of the star shows neither strong CO absorption nor other spectral features in that wavelength region.

4.4 Imaging[3]

To derive the black hole mass in NGC 4486a, it is essential to determine the gravitational potential made up by the stellar component by deprojecting the surface brightness distribution. As NGC 4486a consists both kinematically and photometrically of two components with possibly different mass-to-light ratios Υ, we deproject bulge and disc separately.

To decompose the two components, we considered the *HST* images in the broad-band F850LP filter, with two Advanced Camera for Surveys (ACS) Wide Field Channel (WFC) pointings of 560 seconds exposure each. The two dithers have no shift in spatial coordinates. The data were reduced by the ST-ECF On-The-Fly Recalibration system, see http://archive.eso.org/archive/hst for detailed information.

Moreover, we use the galfit package (Peng et al., 2002) to fit PSF convolved analytic profiles to the two-dimensional surface brightness of the galaxy. The code determines the best fit by comparing the convolved models with the science data using a Levenberg-Marquardt downhill gradient algorithm to minimize the χ^2 of the fit. The saturated star close to the galaxy centre has been masked out from the modelling. The observing strategy, i.e. the adopted no spatial shift between the two dithers, has allowed us to obtain a careful description of the PSF by using the TinyTim[4] code.

---

[3] Based on observations made with the Advanced Camera for Surveys onboard the NASA/ESA *Hubble Space Telescope* (GO Proposal 9401, obtained from the ESO/ST-ECF Science Archive Facility)

[4] http://www.stsci.edu/software/tinytim/tinytim

CHAPTER 4. THE SUPERMASSIVE BLACK HOLE OF NGC 4486A

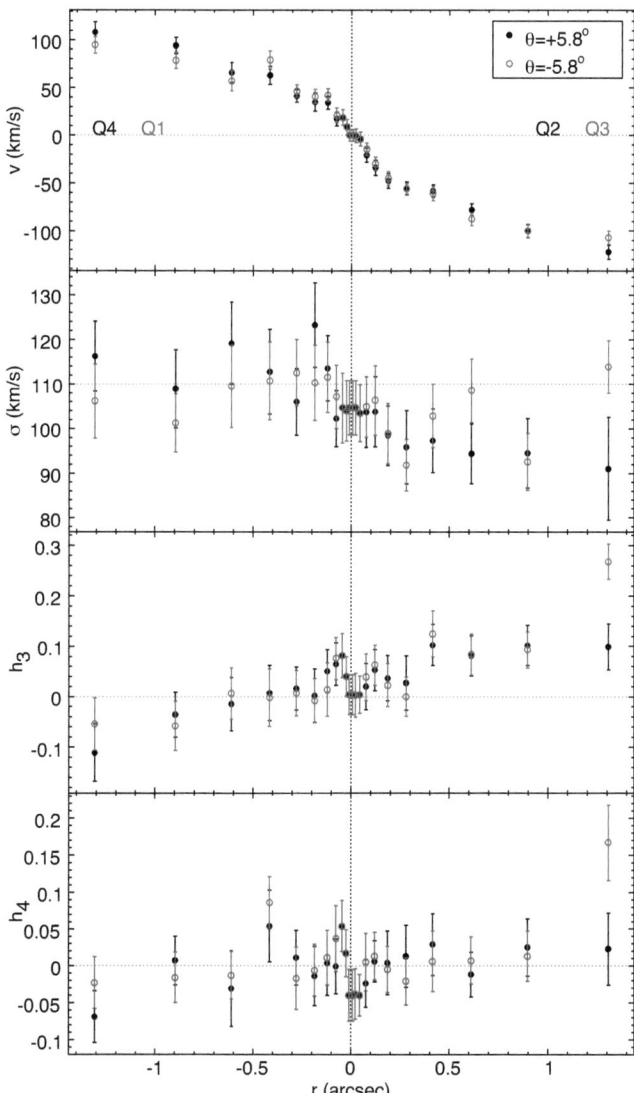

Figure 4.4: Stellar kinematic profiles (v, σ, h_3, h_4) of NGC 4486a along $\theta = \pm 5.8°$ of the major axis. The corresponding quadrants crossed by the profiles are marked for easy identification in Fig. 4.3.

4.4. IMAGING

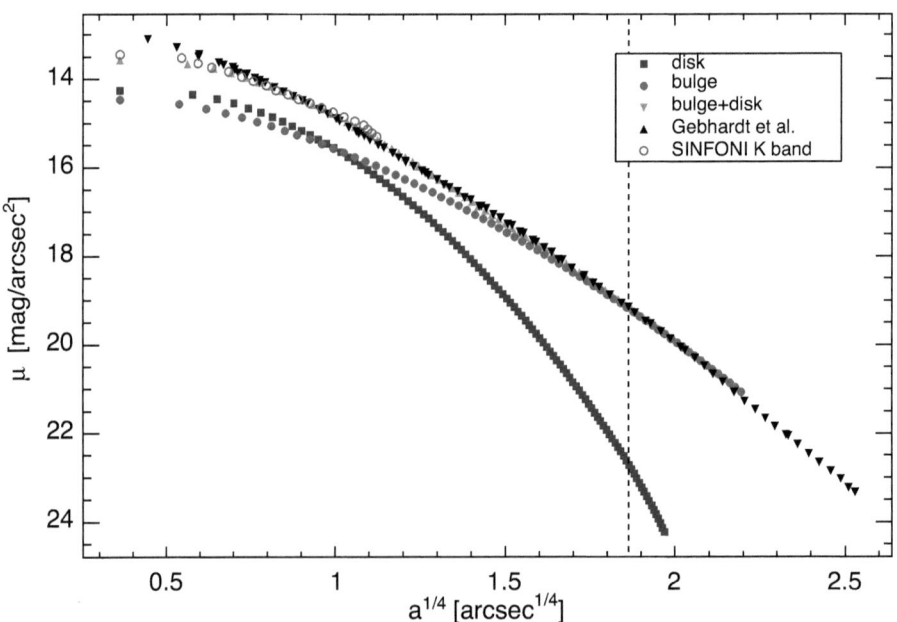

Figure 4.5: Surface brightness profile for NGC 4486a. At radii left of the dashed line we used the decomposed bulge and disc profiles for the deprojection, right of the dashed line we used the profile of Gebhardt et al. (in preparation) (scaled to ACS z-band).

We modelled the galaxy light with a double Sérsic (1968) model with indices $n = 2.19$ for the bulge and $n = 1.67$ for the disc. In Fig. 4.5 we show the surface brightness profiles of bulge, disc, bulge+disc, and for comparison the SINFONI surface brightness profile and the un-decomposed profile of Gebhardt et al. (in preparation), which was derived from *HST* and CFHT imaging in several bands. This profile agrees very well with our combined profile and we use it, scaled to match our profile, at radii $> 11.5''$ where the bulge strongly dominates.

Bulge and disc were then deprojected separately using the code of Magorrian (1999) under the assumption that both components are edge-on and axisymmetric. The stellar mass density then can be modelled as in Davies et al. (2006) via $\rho_* = \Upsilon_{\text{bulge}} \nu_{\text{bulge}} + \Upsilon_{\text{disc}} \nu_{\text{disc}}$, where ν is the luminosity density obtained from the deprojection and the ACS z-band mass-to-light ratio Υ is assumed to be constant with radius for both components.

4.5 Schwarzschild modelling

The mass of the black hole in NGC 4486a was determined based on the Schwarzschild (1979) orbit superposition technique, using the code of Gebhardt et al. (2000a, 2003) in the version of Thomas et al. (2004). It comprises the usual steps: (1) Calculation of a potential with a trial black hole of mass $M_\bullet$ and a stellar mass density ρ_*. (2) A representative set of orbits is run in this potential and an orbit superposition that best matches the observational constraints is constructed. (3) Repetition of the first two steps with different values for Υ_{bulge}, Υ_{disc} and $M_\bullet$ until the eligible parameter space is systematically sampled. The best-fitting parameters then follow from a χ^2-analysis.

The models are calculated on the grid with 11 radial and 5 angular bins per quadrant as described above (cf. Fig. 4.3).

Our orbit libraries contain 2×7000 orbits. The luminosity density is a boundary condition and hence exactly reproduced. The 11×5 LOSVDs are binned into 17 velocity bins each and then fitted directly, not the parametrized moments. We limit the parameter space for the values of Υ by considering the population synthesis model of Maraston (1998, 2005), which gives us $\Upsilon \lesssim 5$ for the z-band.

Special care was taken when implementing the PSF. Due to its special shape with the narrow core and the broad wings (cf. Fig. 4.2) the PSF was not fitted, rather the two-dimensional image of the star was directly used for convolving our models.

4.6. RESULTS

Table 4.1: Results obtained for the four quadrants separately with a global mass-to-light ratio (90% C.L.). The numbering of the quadrants can be inferred from Fig. 4.3.

Quadrant	$M_\bullet$ (10^7 $M_\odot$)	Υ
1	$4.0^{+0.7}_{-2.4}$	$3.4^{+0.6}_{-0.6}$
2	$1.0^{+1.1}_{-0.2}$	$3.6^{+0.2}_{-0.5}$
3	$1.0^{+0.5}_{-0.5}$	$3.8^{+0.6}_{-0.1}$
4	$1.5^{+1.0}_{-1.0}$	$4.4^{+0.3}_{-0.9}$
1 – 4 averaged	$1.25^{+0.3}_{-0.3}$	$4.0^{+0.1}_{-0.4}$

4.6 Results

A big advantage of integral-field data compared to longslit data is that we can check the assumption of axisymmetry by comparing the kinematics of the four quadrants and quantify the effect of possible deviations by modelling each quadrant separately. We do not find major differences in the kinematics of the four quadrants (cf. Figs. 4.3, 4.4). As it takes a large amount of computing time to calculate all models with different mass-to-light ratios for bulge and disc for each quadrant, we used the same Υ for bulge and disc for the comparison of the quadrants. In Table 4.1 the resulting values for $M_\bullet$ and Υ are listed. They show that the four quadrants agree reasonably well with each other. The only systematically deviant point is the $M_\bullet$ determination in the first quadrant that, however, has a large error and therefore is compatible within the 90% confidence limit with the other three quadrants. Therefore we symmetrised the LOSVDs by taking for each bin the weight-averaged LOSVDs of the four quadrants and the corresponding errors. The results of modelling these averaged LOSVDs are shown in Fig. 4.6, where $\Delta\chi_0^2 = \chi^2 - \chi^2_{\min}$ is plotted as a function of $\left(\Upsilon_{\text{bulge}} + \Upsilon_{\text{disc}}\right)/2$ and $M_\bullet$ with error contours for two degrees of freedom. Υ_{bulge} and Υ_{disc} anticorrelate such that their sum is approximately constant, as shown in the upper part of Fig. 4.6. A black hole mass of $(1.25^{+0.75}_{-0.79}) \times 10^7$ $M_\odot$ (90% C.L.) can be fitted with $\Upsilon_{\text{disc}} \approx 2.8\ldots5.2$ and $\Upsilon_{\text{bulge}} \approx 2.8\ldots4.8$. This result agrees within 90% confidence limit with the results of all quadrants shown in Table 4.1. The best-fitting model, obtained with minimal regularisation, is marked with a white circle and has a black hole mass $M_\bullet = 1.25 \times 10^7 M_\odot$ and mass-to-light ratios $\Upsilon_{\text{disc}} = 4.0$ and $\Upsilon_{\text{bulge}} = 3.6$. The difference in χ^2 to the best-fitting model without black hole is 24.14 which corresponds to 4.5 σ. The total χ^2 values for the models are around 300. Together with the number of observables (11 radial bins ×5 angular bins ×17 velocity bins) this gives a reduced χ^2 of ≈ 0.3. Note, however, that the number of observables is in reality smaller due to the smoothing (Gebhardt et al., 2000a).

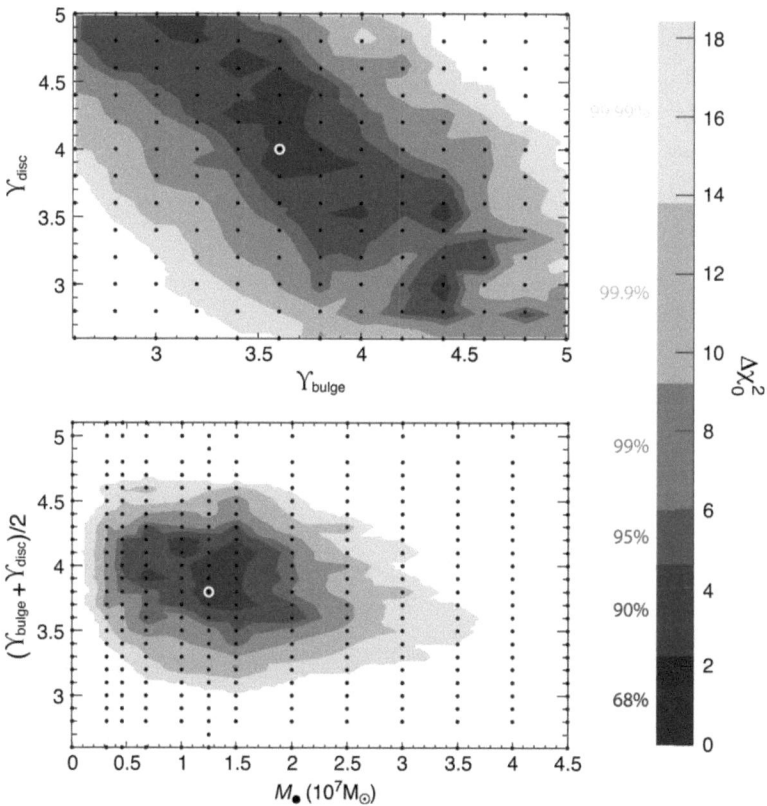

Figure 4.6: $\Delta\chi_0^2 = \chi^2 - \chi^2_{min}$ as a function of (top) Υ_{bulge} and Υ_{disc}, minimized over $M_\bullet$; (bottom) $M_\bullet$ and $\left(\Upsilon_{bulge} + \Upsilon_{disc}\right)/2$. The black points are the models we calculated and the coloured regions are the (unsmoothed) confidence intervals for two degrees of freedom. The best-fitting model is marked with a white circle.

4.6. RESULTS

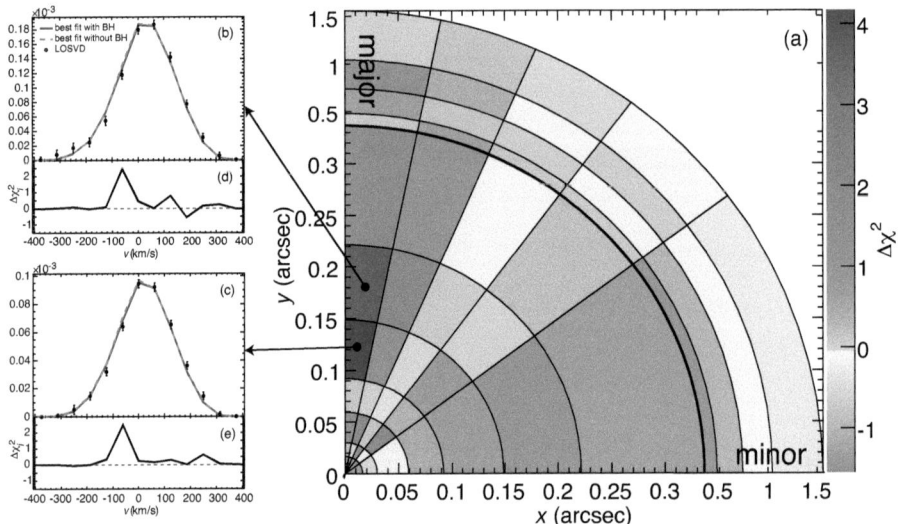

Figure 4.7: (a) χ^2 difference between the best-fitting model without black hole and the best-fitting model with black hole ($\Delta\chi^2 = \sum_i \Delta\chi_i^2 = \sum_{i=1}^{17} \left(\chi^2_{i,\text{noBH}} - \chi^2_{i,\text{BH}}\right)$ over all 17 velocity bins) for all LOSVDs of the averaged quadrant. Bins where the model with black hole fits the LOSVD better are plotted in green, the others in orange. The part outside ≈ 2.5 spheres of influence is plotted with a different scale than the inner part, since in the outer region the dynamical effect of the black hole (and therefore the difference between the two models) is negligible. (b,c) For the radii with the largest positive χ^2 difference in (a) the LOSVD (open circles with error bars, normalised as in Gebhardt et al. 2000a) and both fits (with black hole, full green line, and without black hole, dashed orange line) are shown with the corresponding $\Delta\chi_i^2$ plotted below (d,e).

The dynamical mass-to-light ratios of disc and bulge agree with an old and metal-rich stellar population (Maraston, 1998, 2005). Υ_{disc} tends to be larger than Υ_{bulge} which is probably due to the presence of dust in the disc. To estimate the effect of the dust on the mass-to-light ratio of the disc we are using the model of Pierini et al. (2004). For the *HST*-F850LP filter and an assumed typical optical depth $\tau \lesssim 0.5$ we obtain an attenuation of $A_\lambda \lesssim 0.36$ mag. This translates the best-fitting $\Upsilon_{\text{disc}} = 4.0$ to a significantly smaller dust-corrected value of $\gtrsim 2.9$. Following the models of Maraston (1998, 2005) this is in good agreement with an estimated $\gtrsim 2$ Gyr younger disc (Kormendy et al., 2005).

The significance of the result is illustrated in Fig. 4.7. It shows the χ^2 difference between the best-fitting model without black hole and the best-fitting model with black hole ($\Delta\chi^2 = \sum_{i=1}^{17}\left(\chi^2_{i,\text{noBH}} - \chi^2_{i,\text{BH}}\right)$ over all 17 velocity bins) for all LOSVDs of the averaged quadrant. The part outside ≈ 2.5 spheres of influence, where the dynamical effect of the black hole is negligible, is displayed in a compressed way in order to emphasize the important inner part. For 75% of all bins the model with black hole produces a fit to the LOSVD better than the model without black hole. The signature of the black hole is imprinted mainly along the major axis, where the largest positive $\Delta\chi^2$ are found at radii $r \approx 0.09''\ldots 0.22''$. At radii $r \lesssim 0.09''$ the differences between both models are smaller due to the effect of the PSF. For the radii with the largest $\Delta\chi^2$ along the major axis the LOSVD and the fits with and without black hole are shown in the left part of Fig. 4.7 together with the corresponding $\Delta\chi_i^2$ as a function of the line-of-sight velocity. The differences between the two fits are relatively small in absolute terms. However, the model without black hole has more stars on the low-velocity wing at a $\sim 1.5\sigma$ level, failing to match fully the measured slightly higher mean velocity of the galaxy. Future observations with a higher spatial resolution should be able to probe this difference more clearly.

The total stellar mass within the sphere of influence, where the imprint of the black hole is strongest, is $M_* = 9.84 \times 10^6\ M_\odot$. If the additional mass of $M_\bullet = 1.25 \times 10^7\ M_\odot$ was solely composed of stars, this would increase the mass-to-light ratio to $\Upsilon_{\text{disc}} \approx 9.1$ (6.6 if we take into account the dust-absorption), a region which is excluded by stellar population models or at least requires unrealistic high stellar ages.

The models with black hole become tangentially anisotropic in the centre, while the models without black hole are close to isotropic.

Our result is in good agreement with the prediction of the $M_\bullet$-σ relation ($1.24 \times 10^7\ M_\odot$ using the result of Tremaine et al. 2002 and the average velocity dispersion of 110.5 km s^{-1} in the 100mas SINFONI field of view) and strengthens it in the low-σ regime ($\lesssim 120$ km s^{-1}), where, besides several upper limits, up to now only four black hole masses were measured with stellar kinematics (Milky Way, Gillessen et al. 2009; M32, Verolme et al. 2002; NGC 4258, Siopis et al. 2009; NGC 7457, Gebhardt et al. 2003).

Acknowledgments

We are grateful to Frank Eisenhauer and Stefan Gillessen for assistance in using the SINFONI instrument and the reduction software spred. Furthermore we thank David Fisher for providing us the surface brightness profile which we used at large radii.

5

The supermassive black hole of Fornax A[1]

Abstract

The radio galaxy Fornax A (NGC 1316) is a prominent merger remnant in the outskirts of the Fornax cluster. Its giant radio lobes suggest the presence of a powerful AGN and thus a central supermassive black hole (SMBH). Fornax A now seems to be in a transition state between active black hole growth and quiescence, as indicated by the strongly declined activity of the nucleus. Studying objects in this evolutionary phase is particularly important in order to understand the link between bulge formation and black hole growth, which is manifested in the $M_\bullet$-σ relation between black hole mass and bulge velocity dispersion. So far a measurement of the SMBH mass has not been possible in Fornax A, as it is enshrouded in dust, which makes optical measurements impossible. We present high-resolution adaptive optics assisted integral-field data of Fornax A, taken with SINFONI at the Very Large Telescope in the K-band, where the influence of dust is negligible. The achieved spatial resolution is $0.085''$, which is about a fifth of the diameter of the expected sphere of influence of the black hole. The stellar kinematics was measured using the region around the CO bandheads at 2.3 μm. Fornax A does not rotate inside the inner $\sim 3''$. The velocity dispersion increases towards the centre. The weak AGN emission affects the stellar kinematics in the inner $\sim 0.06''$ only. Beyond this radius, the stellar kinematics

---
[1]This chapter has been published as "Nowak N., Saglia R. P., Thomas J., Bender R., Davies R. I., Gebhardt K., 2008, MNRAS, 391, 1629"

appears relaxed in the central regions. We use axisymmetric orbit models to determine the mass of the SMBH in the centre of Fornax A. The three-dimensional nature of our data provides the possibility to directly test the consistency of the data with axisymmetry by modelling each of the four quadrants separately. According to our dynamical models, consistent SMBH masses $M_\bullet$ and dynamical K_s-band mass-to-light ratios Υ are obtained for all quadrants, with $\langle M_\bullet \rangle = 1.3 \times 10^8$ M$_\odot$ (rms($M_\bullet$) = 0.4×10^8 M$_\odot$) and $\langle \Upsilon \rangle = 0.68$ (rms(Υ) = 0.03), confirming the assumption of axisymmetry. For the folded and averaged data we find $M_\bullet = 1.5^{+0.75}_{-0.8} \times 10^8$ M$_\odot$ and $\Upsilon = 0.65^{+0.075}_{-0.05}$ (3 σ errors). Thus the black hole mass of Fornax A is consistent within the error with the Tremaine et al. (2002) $M_\bullet$-σ relation, but is a factor ~ 4 smaller than expected from its bulge mass and the Marconi & Hunt (2003) relation.

5.1 Introduction

Studies of the dynamics of stars and gas in the centres of nearby galaxies with a massive bulge component have established the presence of supermassive black holes (SMBHs) in the $10^6 - 10^9$ M$_\odot$ range. The mass of the central SMBH is tightly correlated with the bulge mass or luminosity (e.g. Marconi & Hunt 2003) and with the bulge velocity dispersion σ (Ferrarese & Merritt, 2000; Gebhardt et al., 2000b). These correlations suggest that bulge evolution and black hole growth are closely linked. Indeed there is increasing theoretical evidence that galaxy merging (or other processes that lead to gas inflow, like secular evolution), nuclear activity and feedback are all somehow linked to bulge and SMBH evolution (e.g. Di Matteo et al. 2005; Hopkins et al. 2008; Johansson et al. 2009; Younger et al. 2008). To observationally constrain present theories of bulge and black hole evolution, detailed studies of AGN and merger remnants in different evolutionary stages are essential. NGC 5128 (Cen A) is presently the only galaxy with a powerful AGN that underwent a recent major merger which has a measured black hole mass (Cappellari et al., 2009; Häring-Neumayer et al., 2006; Krajnović et al., 2007; Marconi et al., 2001, 2006; Neumayer et al., 2007; Silge et al., 2005).

A galaxy very similar to Cen A is NGC 1316 (Fornax A). Fornax A is a giant elliptical galaxy located in the outskirts of the Fornax galaxy cluster. It is one of the brightest radio galaxies in the sky with giant double radio lobes and S-shaped nuclear radio jets (Geldzahler & Fomalont, 1984). The peculiar morphology with numerous tidal tails, shells and loops (Schweizer, 1980, 1981) suggests that a major merger happened about 3 Gyr ago (Goudfrooij et al., 2001), followed by some minor mergers (Mackie & Fabbiano, 1998). Its nucleus, however, is surprisingly faint in X-rays, which can only be explained if the nucleus became dormant during the last 0.1 Gyr (Iyomoto et al., 1998). This makes it an ideal target for a dynamical black hole mass measurement, as its stellar absorption lines are probably not or just marginally diluted by non-stellar emission from the AGN.

Fornax A is classified as an intermediate form between core and power law galaxy in Lauer et al. (2007). The surface brightness approximately follows an $r^{1/4}$-law (Caon et al., 1994). Large amounts of dust are present also in the inner few arcseconds, which affects optical spectroscopy. We therefore measure the two-dimensional stellar kinematics of Fornax A with the near-infrared integral-field spectrograph SINFONI at the Very Large Telescope (VLT) in the K-band, where the dust obscuration is only about 7% of the obscuration in the optical. We use this data to analyse the stellar kinematics in a way similar to Chapter 4 (Nowak et al., 2007).

Throughout this Chapter we adopt a distance to Fornax A of 18.6 Mpc based on *HST* measurements of Cepheid variables in NGC 1365 (Madore et al., 1999). At this distance, $1''$ corresponds to 90 pc. With a velocity dispersion of 226 km s^{-1} (mean σ measured in an $8''$ aperture in Section 5.3.4) the estimated sphere of influence has a diameter of $0.46''$ and we would expect a SMBH mass of 2.2×10^8 M$_\odot$ (Tremaine et al., 2002). This is large enough to be resolved easily from the ground with adaptive optics.

This Chapter is organized as follows: In Section 5.2 we present the data and the data reduction. The stellar kinematics of Fornax A is described in Section 5.3, the measurement of the near-infrared line strength indices in Section 5.4. The stellar dynamical modelling procedure and the results for the SMBH mass are presented in Section 5.5, and Section 5.6 summarises and discusses the results.

5.2 Data and data reduction[2]

5.2.1 SINFONI data

Fornax A was observed between 2005 October 10 and 12 as part of guaranteed time observations with SINFONI (Bonnet et al., 2004; Eisenhauer et al., 2003a), an adaptive-optics assisted integral-field spectrograph at the VLT UT4. The nucleus of Fornax A served as guide star for the AO correction. We used the K-band grating (1.95 – 2.45 µm) and, depending on the seeing conditions, the intermediate size field of view of $3 \times 3''$ ($0.05 \times 0.1''$/spaxel, referred to as "100mas scale" in the following) or the high resolution mode with a $0.8 \times 0.8''$ field of view ($0.0125 \times 0.025''$/spaxel, "25mas scale"). The total on-source exposure time was 130 min in the highest resolution mode

---

[2]Based on observations at the European Southern Observatory (ESO) Very Large Telescope (VLT) [076.B-0457(A)] and archival ESO La Silla [066.C-0310(A)] and NASA/ESA *Hubble Space Telescope* data (GO Proposal 7458), obtained from the ESO/ST-ECF Science Archive Facility.

5.2. DATA AND DATA REDUCTION

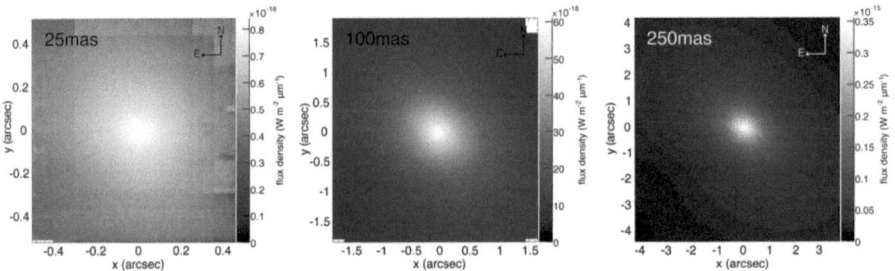

Figure 5.1: SINFONI images of Fornax A in the 25mas (left), 100mas (middle) and 250mas (right) scale.

and 70 min using the intermediate plate scale, consisting of 10 min exposures taken in series of "object-sky-object" (O-S-O) cycles, dithered by up to $0.2''$. In addition three O-S-O cycles of 5 min exposure time each were taken using the low resolution mode ($8 \times 8''$ field of view, "250mas scale").

The SINFONI data reduction package SPRED (Abuter et al., 2006; Schreiber et al., 2004) was used to reduce the data. It includes all common reduction steps necessary for near-infrared (near-IR) data plus routines to reconstruct the three-dimensional datacubes. The data were skysubtracted, flatfielded, corrected for bad pixels and for distortion and then wavelength calibrated using a Ne/Ar lamp frame. The wavelength calibration was corrected using night-sky lines if necessary. Then the datacubes were reconstructed and corrected for atmospheric absorption using B stars that do not have strong spectral features in the region of the CO bandheads. As a final step all datacubes were averaged together to produce the final datacube. The data of the telluric stars were reduced likewise. The flux calibration of the 250mas data was performed by comparison to the photometrically calibrated NTT/SOFI K_s-band imaging (see Section 5.2.3 and Fig. 5.3), that yields a K_s magnitude of 8.44 in an $8''$ aperture. The 100mas image was calibrated with respect to the 250mas image and the 25mas image with respect to the 100mas image. Fig. 5.1 shows the resulting image of Fornax A in all three platescales, collapsed along the wavelength direction within the SOFI K_s-band region.

The spatial resolution, i.e. the FWHM of the point-spread function (PSF), is difficult to derive when dealing with high-resolution AO observations of diffuse objects. Deriving it from the science data itself can work quite well, if there are either stars in the field of view or the target is an AGN or starburst galaxy with certain spectral properties (see e.g. Davies et al. 2006, 2007; Mueller Sánchez et al. 2006). None of this applies to Fornax A. Therefore we derived the PSF by regularly taking an exposure of a nearby star with approximately the same R-band magnitude and $B - R$ colour as the central $3''$ of Fornax A. This procedure is usually recommended to observers, but apart from being very time-consuming it can also be unreliable under certain conditions (Davies et al., 2004).

The atmosphere may vary strongly with time, resulting in a measured PSF that is different from the real one. In addition the response of the wavefront sensor to the PSF star is, to some degree, different from the response to the AO guide star (the nucleus in this case), because the nucleus is extended and therefore the flux distribution on the wavefront sensor is different. We used this method as it was the only available option to measure the PSF.

The high-resolution (25mas) data were taken under good and relatively stable ambient conditions with a near-IR seeing around 0.6″. The two-dimensional PSF (i.e. the image of the PSF reference star) is shown in Fig. 5.2a. The spatial resolution is FWHM $\approx$ 0.085″. A Strehl ratio of $\approx$ 0.45 was reached. The reliability of this PSF is verified by a comparison of the 25mas SINFONI luminosity profile with an *HST*/NICMOS F160W (camera NIC2, sampled at 0.075″/px) luminosity profile. They agree well without further broadening of the NICMOS data (see below and Fig. 5.4).

During the hours of degraded seeing (around $\approx$ 0.9″ on average in the near-IR) on October 12 we observed Fornax A using the intermediate plate scale (100mas). The spatial resolution inferred from the PSF is $\approx$ 0.16″ and the achieved Strehl ratio was $\approx$ 0.18. The shape of the PSF (Fig. 5.2b) is rather asymmetric and may not represent the true PSF due to the changeable weather conditions. The seeing and the coherence time drastically improved during the observations with the 100mas scale, so we switched to the 25mas scale for approximately one hour and only when the seeing and the coherence time then degraded again to similar values as before we could observe the 100mas PSF. In order to get a better estimate of the real PSF we computed the kernel that transforms the 25mas image of Fornax A (binned 4 × 4 to match the 100mas spaxel size) into the 100mas image using the code of Gössl & Riffeser (2002). The 25mas PSF was then rebinned to the 100mas spaxel size and convolved with this kernel. Kernel and resulting PSF are shown in Fig. 5.2c and 5.2d. The shape of the kernel and hence the convolved PSF are quite noisy. The reason is the low S/N of the 25mas image compared to the 100mas image (cf. Fig. 5.1). In order to find the closest match between the two images the kernel tries to incorporate the noise, too. Thus the resulting PSF has a S/N more similar to the 25mas image than to the image of a bright PSF star. Nevertheless the convolved PSF has about the same FWHM as the measured 100mas PSF star.

The PSF is an integral part of the dynamical modelling and it is therefore important to know its shape as accurate as possible. However, so far no studies have been performed that analyse systematically the dependence of the resulting $M_\bullet$ and mass-to-light ratio on the PSF shape. Such a study is beyond the scope of this thesis, nevertheless we will do the modelling, as far as 100mas data are included, twice, using one time the measured PSF and the other time the convolved PSF.

5.2. DATA AND DATA REDUCTION

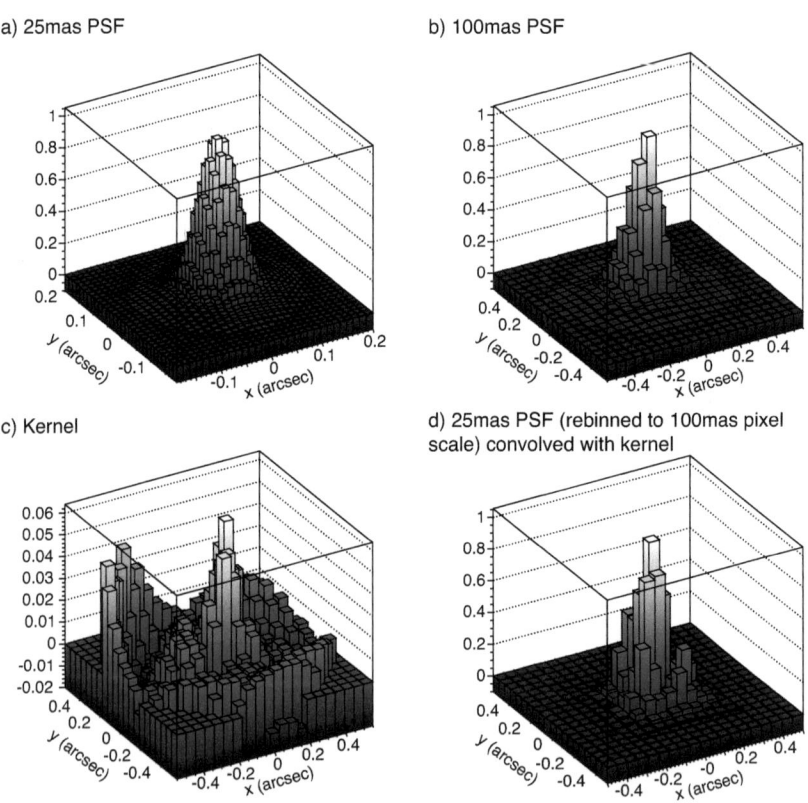

Figure 5.2: Panel (a): 25mas PSF derived by observing a star of about the same magnitude and colour as the AO guide star (i.e. the nucleus of Fornax A). Panel (b): Same as panel (a) but for the 100mas scale. Panel (c): Kernel that transforms the 25mas image of Fornax A to the 100mas image. Panel (d): 25mas PSF from panel (a) rebinned to the 100mas spaxel size and convolved with the kernel from panel (c).

5.2.2 Longslit data

Kinematics derived from longslit spectra are useful to constrain the orbital structure at large radii, outside the small SINFONI field of view. Major-axis longslit data of Fornax A are available from different authors (Arnaboldi et al., 1998; Bedregal et al., 2006; Longhetti et al., 1998; Saglia et al., 2002). Arnaboldi et al. (1998) and Longhetti et al. (1998) measured only v and σ from optical absorption lines (Mgb) and from the CaT region in the near-IR, they did not measure the higher-order Gauss-Hermite coefficients h_3 and h_4 (Gerhard, 1993; van der Marel & Franx, 1993) which quantify asymmetric and symmetric deviations from a Gaussian velocity profile. Bedregal et al. (2006) measured all four parameters from optical spectra, whereas Saglia et al. (2002) used the CaT region. The velocities of all measurements are consistent, taking into account the different seeing values. The velocity dispersions in the central $\sim 5''$, however, range from ≈ 220 km s^{-1} to ≈ 260 km s^{-1}. The reasons for that can be diverse. The authors used different correlation techniques, different spectral lines and slightly different position angles. Fornax A contains dust, which might alter the kinematics measured from optical absorption lines. Silge & Gebhardt (2003) found that optical dispersions tend to be larger than those measured from the CO bandheads, which could be an effect of dust. The largest dispersions are indeed those measured from optical spectra. We therefore used the kinematics of Saglia et al. (2002), as they measured all four parameters from the CaT line region. Their measurements agree best with our SINFONI measurements. They observed Fornax A in October 2001 at the Siding Spring 2.3m telescope using a $6.7' \times 4''$ longslit and an exposure time of 60 min and determined the kinematics with the Fourier Correlation Quotient method (Bender et al., 1994). As they assume a position angle of 58° we adopt this value in the following analysis. The major axis profiles are shown in Fig. 5.17 and in Beletsky et al. (2009).

5.2.3 Imaging

To measure the SMBH mass in Fornax A, it is essential to determine the gravitational potential made up by the stellar component using photometric measurements with sufficient spatial resolution and radial extent. Therefore we combine high-resolution *HST* NICMOS F160W and ground-based wide-field K_s-band imaging. The low-resolution K_s-band image was taken with the near-IR imager/spectrometer SOFI on the 3.5m NTT telescope on La Silla, Chile. It has a field of view of $4.9 \times 4.9'$ with $0.29''$/px and a seeing of $\sim 0.7''$. It was dust corrected with the method described below using a *J*-band SOFI image. The dust corrected image is shown in Fig. 5.3. Note that the original SOFI image and the residuals are shown in Beletsky et al. (2009). We applied the same

5.2. DATA AND DATA REDUCTION

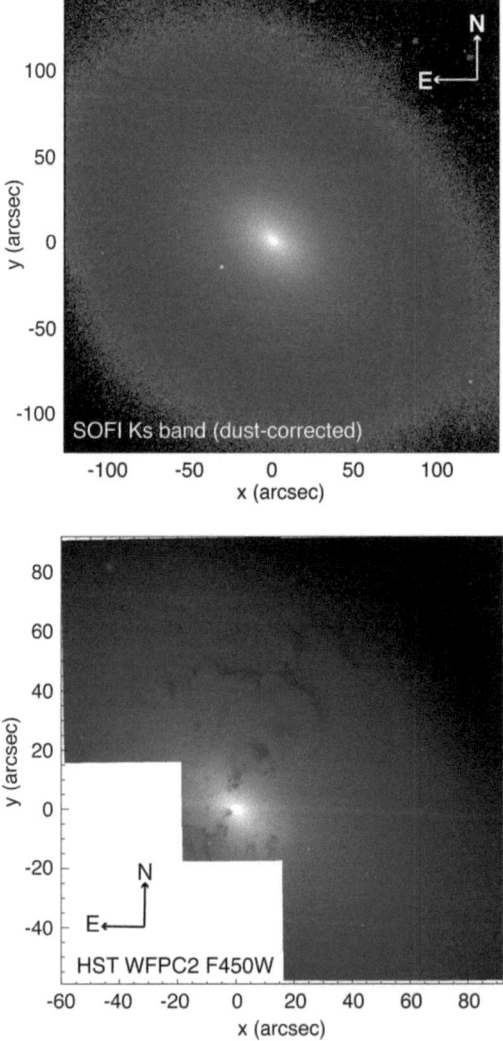

Figure 5.3: The top panel shows the dust corrected SOFI K_s-band image of Fornax A (courtesy Y. Beletsky). The dust corrected light distribution appears very smooth and regular and the isophotes are slightly boxy. For comparison the *HST* WFPC2 F450W image (NASA/ESA *HST* data (GO Proposal 5990), obtained from the ESO/ST-ECF Science Archive Facility), which highlights the strong dust lanes, is presented in the bottom panel.

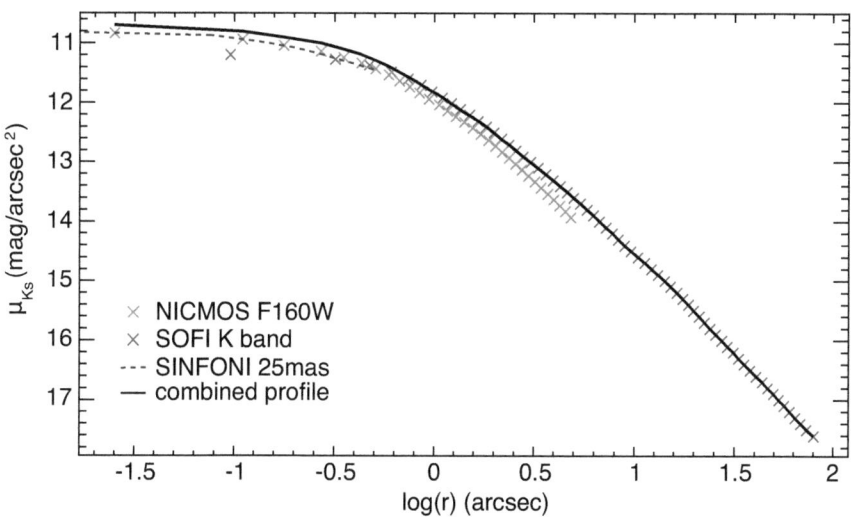

Figure 5.4: Surface brightness profiles of the *HST* NICMOS dust corrected *H*-band image (green crosses), the SOFI dust corrected K_s-band image (red crosses) and the combination of the two surface brightness profiles (black line). The SINFONI 25mas *K*-band image (scaled to match the *HST* NICMOS profile) is overplotted in blue.

method to the *HST* NICMOS F160W image, which we corrected using a NICMOS F110W image. The isophotal profiles of the dust corrected images were obtained following Bender & Moellenhoff (1987). The light distribution appears very smooth and regular and the isophotes are slightly boxy. The surface brightness profiles were then combined using the method of Corsini et al. (2008), matching the profiles in the region between 1.5″ and 5.0″ and shifting the NICMOS profile by the computed amount ($\mu_{\text{SOFI}} - \mu_{HST}$) on to the K_s-band SOFI profile. The original and the combined profiles are shown in Fig. 5.4.

The combined photometry extends out to 80″ and was deprojected for an inclination $i = 90°$ using the code of Magorrian (1999) under the assumption that the galaxy is axisymmetric. No shape penalty was applied. The stellar mass density then can be modelled via $\rho_* = \Upsilon \nu$, where ν is the luminosity density obtained from the deprojection. The K_s-band mass-to-light ratio Υ is assumed to be constant with radius. This assumption holds approximately in the central part of the galaxy for which we have kinematic data and where dark matter does not play a major role.

5.3. STELLAR KINEMATICS

Dust correction

The ground-based K_s-band SOFI and the H-band *HST* NICMOS imaging data were dust corrected with according J-band images taken with the same instruments using the following method, assuming a standard extinction law: The relation between reddening and extinction is

$$H_0 - H_c = a E_{J-H} \tag{5.1}$$

for the *HST* image (analogous for the SOFI image), where H_0 is the observed, and H_c is the dust-free image. The absorption coefficient a is defined via the extinctions in the involved bands: $a = A_{H,K}/(A_J - A_{H,K})$. For the SOFI image the values $A_J = 0.282 A_V$ and $A_K = 0.112 A_V$ from Binney & Merrifield (1998) were used. The NICMOS image was significantly overcorrected with A_J and A_H from Binney & Merrifield (1998). We found by iteratively changing a that the best correction is obtained for $a = 0.8$.

The observed images provide the fluxes, so with Eq. 5.1 it follows that

$$H_c = H_0 - a[(J-H)_0 - (J-H)_c] \tag{5.2}$$

and

$$f_{H_c} = f_{H_0} \left(\frac{f_{H_0}}{f_{J_0}} \right)^a \left(\frac{f_{J_c}}{f_{H_c}} \right)^a \propto \frac{f_{H_0}^{1+a}}{f_{J_0}^a} \tag{5.3}$$

under the assumption that $f_{J_c}/f_{H_c} \approx$ const. (i.e. approximately no intrinsic stellar population gradient). See also Beletsky et al. (2009).

5.3 Stellar kinematics

The stellar kinematic information are extracted using the maximum penalised likelihood (MPL) technique of Gebhardt et al. (2000a), which obtains non-parametric line-of-sight velocity distributions (LOSVDs) as follows: an initial binned velocity profile is convolved with a linear combination of template spectra and the residuals of the resulting spectrum to the observed galaxy spectrum are calculated. The velocity profile is then changed successively and the weights of the templates are adjusted in order to optimise the fit to the observed spectrum by minimizing the function $\chi_P^2 = \chi^2 + \alpha \mathscr{P}$, where α is the smoothing parameter that determines the level of regularisation, and the penalty function $\mathscr{P}$ is the integral of the square of the second derivative of the LOSVD. We fitted only the first two bandheads $^{12}CO(2-0)$ and $^{12}CO(3-1)$, i.e. the spectral range between

2.275 μm and 2.349 μm rest frame wavelength. The higher-order bandheads are strongly disturbed by residual atmospheric features. At wavelengths $\lambda < 2.29$ μm most absorption lines are weak such that extremely high S/N spectra would be needed. The strongest absorption line in that regime is Na I, but as its strength is much higher in early-type galaxies than in single stars due to enhanced silicon (Silva et al., 2008) it cannot be fitted by the template spectra.

5.3.1 Initial parameters

Before being able to obtain the final LOSVDs it is necessary to quantify possible systematic offsets and to constrain the best initial values for the smoothing parameter and the width of the LOSVD bins for the given data set (wavelength range, σ, h_3, h_4, S/N, spectral resolution). Especially the selection of the smoothing parameter α is a crucial step (Joseph et al., 2001; Merritt, 1997). If α is chosen too high, the LOSVD is biased toward a flatter shape and if it is too small the LOSVD is too noisy. In order to find the best initial values for the Fornax A data Monte Carlo simulations on a large set of model galaxy spectra have been performed (see Section 3.4.2).

Based on the simulations we conclude that with MPL reliable LOSVDs can be obtained from the first two CO bandheads of the SINFONI data of Fornax A when the chosen α is $\lesssim 10$ and the S/N$\gtrsim 30$. The S/N of our data is very high (on average ~ 70 for the 25mas and the 250mas data, and ~ 140 for the 100mas data). We are using $\alpha = 8$ for the 25mas and 250mas data and $\alpha = 6$ for the 100mas data.

5.3.2 Kinematic template stars

We built up a small library of late-type (K and M) kinematic template stars which were observed with SINFONI in the K-band during commissioning and our GTO observations between 2004 and 2006. In total we have nine kinematic templates for the 25mas scale (spectral resolving power $R \approx 5950$), twelve templates for the 100mas scale ($R \approx 5090$) and ten templates for the 250mas scale ($R \approx 4490$). As shown in Silge & Gebhardt (2003) a correct value of the velocity dispersion σ can only be obtained from fits to the CO bandheads, if the template has about the same intrinsic CO equivalent width (EW) as the galaxy. The CO EWs of the Fornax A spectra were measured using the definition and the velocity dispersion correction of Silge & Gebhardt (2003). The resulting values are between 13.0 Å and 14.5 Å.

After excluding all stellar templates with EWs far below or above the measured range, five templates remained for the 25mas scale, five for the 100mas scale and four for the 250mas scale. As a cross-check and to avoid template mismatch the kinematics was extracted first with single tem-

5.3. STELLAR KINEMATICS

Table 5.1: CO equivalent widths of the stellar kinematic template stars.

Name	spectral type	CO^a	CO^b	scale
HD12642	K5 III	12.7	17.1	25
HD163755	K5/M0 III	15.6	21.3	25
HD179323	K2 III	11.1	14.1	25
HD198357	K3 III	10.8	13.8	25
HD75022	K2/3 III	12.0	15.1	25, 100
HD141665	M5 III	15.8	22.3	100, 250
HD181109	K5/M0 III	12.9	17.8	100, 250
HD201901	K3 III	11.4	14.4	100, 250
SA112-0595^c	M0 III	12.8	17.61	100, 250

Note. The CO equivalent widths are given in units of Å. The spectral types are taken from Wright et al. (2003) where available. In the last column the SINFONI platescales, in which the stars have been observed, are given.
^{a}Definition of Silge & Gebhardt (2003); ^{b}Definition of Silva et al. (2008); ^{c}N. Neumayer, private communication.

plates. The results agreed, so no other template star had to be excluded from the sample. Tab. 5.1 shows the used kinematic template stars with the according spectral types and CO EWs. For a better comparison with the line strength indices discussed in Section 5.4 the CO EWs were also measured using the definition of Silva et al. (2008).

5.3.3 Error estimation

The uncertainties on the velocity profiles are estimated using Monte Carlo simulations (Gebhardt et al., 2000a). A galaxy spectrum is created by convolving the template spectrum with the measured LOSVD. Then 100 realizations of that initial galaxy spectrum are created by adding appropriate Gaussian noise. The LOSVDs of each realization are determined and used to specify the confidence intervals.

5.3.4 The kinematics of Fornax A

The SINFONI data are binned using a similar binning scheme as in Nowak et al. (2007) with five angular bins per quadrant and a number of radial bins (seven for the 25mas scale, ten for the 100mas scale and 13 for the 250mas scale). The centres of the angular bins are at latitudes $\vartheta = 5.8°$, 17.6°,

CHAPTER 5. THE SUPERMASSIVE BLACK HOLE OF FORNAX A

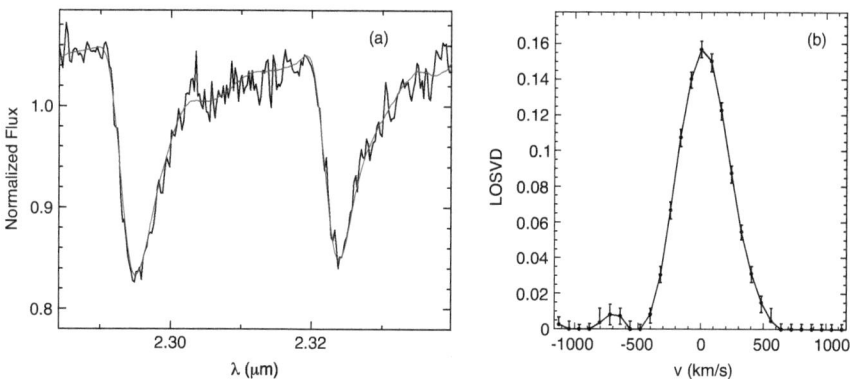

Figure 5.5: a) Example fit to a 25mas spectrum and b) the according LOSVD with error bars.

30.2°, 45.0° and 71.6°. The stellar kinematics then was extracted for all scales using the appropriate templates and α. Fig. 5.5 shows as an example the fit to a 25mas spectrum and the according LOSVD.

For the dynamical modelling we use these LOSVDs and no parametrized moments, but for illustration purposes we fitted the LOSVDs with Gauss-Hermite polynomials and we show the 2D fields of v, σ and the Gauss-Hermite coefficients h_3 and h_4 in Fig. 5.6 for all platescales. The amplitudes of the errors are on average about 8 km s^{-1} (v and σ) or 0.025 (h_3 and h_4) for the 25mas data, 4 km s^{-1} or 0.015 for the 100mas data and 9.5 km s^{-1} or 0.033 for the 250mas data.

On large scales (250mas, Fig. 5.6c) the galaxy is clearly rotating and the mean velocity dispersion is high ($\sigma \approx 226$ km s^{-1}), while in the central $\sim 3''$ (100mas, Fig. 5.6b) no clear rotation is visible and the dispersion is lower ($\sigma \approx 221$ km s^{-1}). The 25mas velocity field (Fig. 5.6a) is irregular and with a peak in the central bins. The velocity dispersion also rises towards the centre, but the maximum does not coincide with the photometric centre. It is located $\approx 0.05 - 0.1''$ south of the centre. North of the centre there is a second but less strong maximum. The two maxima are separated by a narrow, elongated σ-minimum. Outside the central region the mean dispersion of the 25mas field is lower than in the larger fields ($\sigma \approx 218$ km s^{-1}). h_3 and h_4 are on average small and positive.

In addition to the difficulty in explaining the structures of v and σ in the centre, it is also surprising that neither the v nor the σ peak can be seen in the 100mas data. We investigated if this is due to resolution in the following way: The 25mas data were binned 4×4 to get the same spaxel sizes as in the 100mas scale. Then the binned data were convolved with the kernel (Fig.

5.3. STELLAR KINEMATICS

5.2c) that transforms the 25mas SINFONI image of Fornax A into the 100mas image, before being binned according to the radial and angular scheme described above. Fig. 5.7a shows the resulting 2D kinematic maps compared to the inner 0.8″ of the 100mas data (Fig. 5.7b). The structures seen in the 25mas data mostly disappeared after the convolution, and the convolved 25mas kinematics now look very similar to the 100mas kinematics.

The interpretation of the features seen in the unconvolved 25mas data is more difficult. In the following possible explanations are discussed.

Technical aspects

The features detected in the 25mas data, especially the central velocity rise, are not due to template mismatch, as they remain also when other template combinations or single templates are used. Of the nine 25mas templates, two (a K1V and a K3V star with CO EWs $\lesssim 5$ Å) cannot fit the data at all. With each of the other templates the kinematics of Fig. 5.6a can be reproduced at least qualitatively independently of the specific choice of the smoothing parameter.

The results can also be reproduced when fitting slightly different wavelength regions, as long as the CO bandheads are included. Measuring the kinematics from the other absorption lines (Ca I, Fe I and Mg I) does not work because of the weakness of the lines compared to CO. Here a much higher S/N would be required.

The central bins are very small and only cover a limited number of spaxels. Therefore one or two spaxels alone could cause the increase of v, because e.g. of an incorrect sky subtraction or bad pixel correction in the CO bandhead region of these spaxels. In addition, as the S/N of the single spaxels decreases relatively slowly from the centre to the outer parts, the S/N of the central bins is lower (~ 50) than at larger radii (60 − 80), so the errors are larger and the large velocity could be just a statistical fluctuation which would occur at larger radii as well if the S/N was equally low. A single-spaxel effect can be in principle verified or excluded by simply fitting the unbinned spectra. The problem is, that the S/N is around or below 30 which does not allow a reliable measurement of the kinematics. We first used MPL with a large α and then, as a consistency check, fitted simple Gaussians to the unbinned spectra. A central v peak is present in both cases.

A central disc or star cluster

Central velocity dispersion drops are common in spiral galaxies (e.g. Peletier et al. 2007) and usually associated with central discs. Central discs can be formed from gas inflow towards the centre and subsequent star formation (Wozniak et al., 2003). In early-type galaxies σ-drops are rarely found.

CHAPTER 5. THE SUPERMASSIVE BLACK HOLE OF FORNAX A

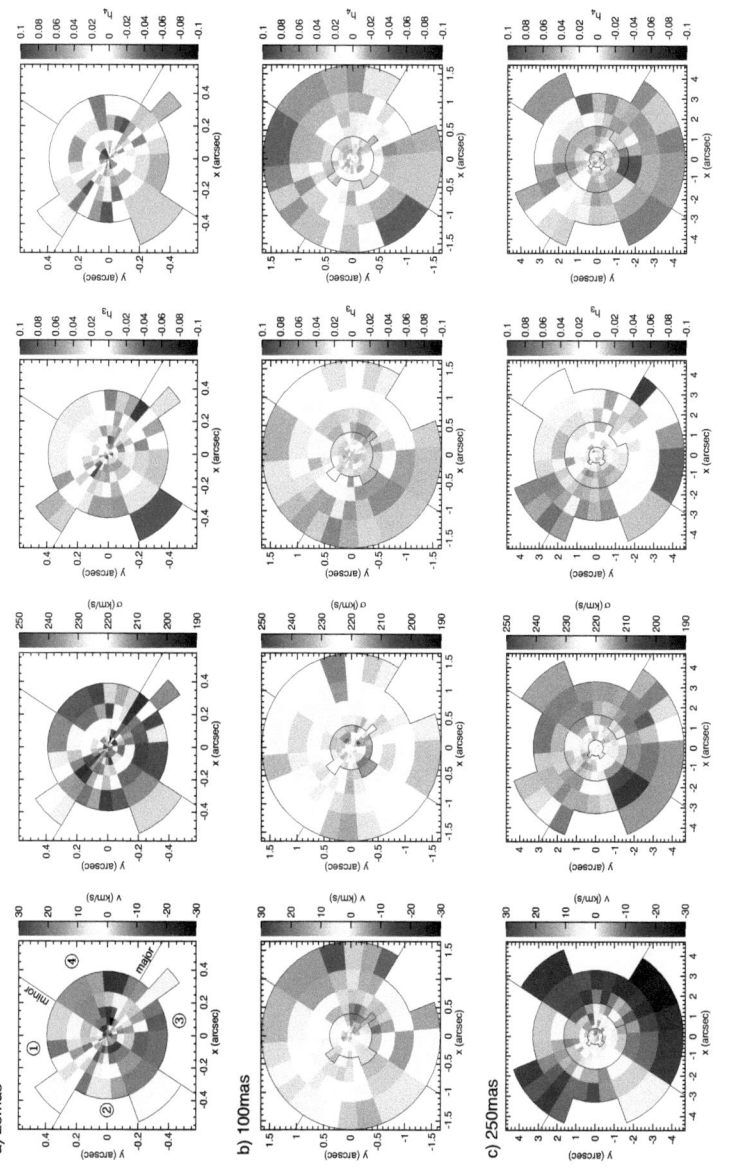

Figure 5.6: Panel (a): v, σ, h_3 and h_4 of Fornax A in the central 0.8″ (25mas scale). The major and the minor axis as well as the numeration of the different quadrants are indicated in the velocity field (left). Panel (b) shows the same as panel (a) but for the 100mas scale. The 25mas field of view is marked for a better orientation. Panel (c) shows the same as panel (a) but for the 250mas scale. Both the 25mas and the 100mas field of view are marked.

5.3. STELLAR KINEMATICS

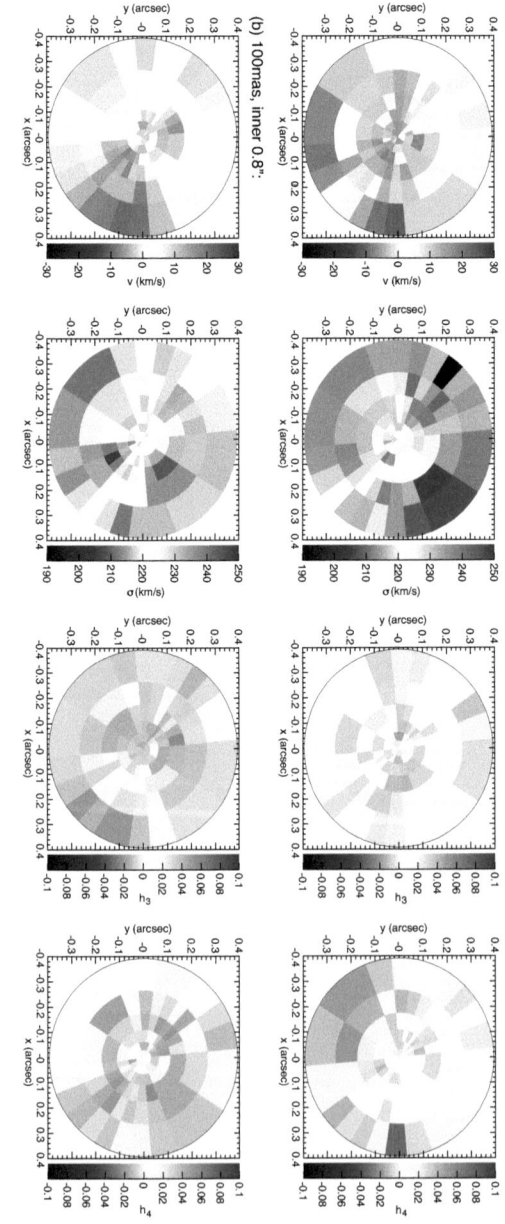

Figure 5.7: The 25mas data convolved with the kernel from Fig. 5.2c (upper row, panel (a)), compared to the central part of the 100mas kinematics (lower row, panel (b)).

NGC 1399 is the only case where the observed central σ-drop is discussed in detail (Gebhardt et al., 2007; Houghton et al., 2006; Lyubenova et al., 2008). Lyubenova et al. (2008) conclude that it is a dynamically distinct subsystem with a different stellar population.

Fornax A is not a clearly elliptical galaxy, but can be classified as a "peculiar" galaxy, as some remaining features like shells and ripples from a major gas-rich merger about 3 Gyr ago are still clearly visible (Goudfrooij et al., 2001; Schweizer, 1980, 1981). There are also hints that a minor merger with another gas-rich galaxy happened ~ 0.5 Gyr ago (Mackie & Fabbiano, 1998). Thus it would not be surprising if remainders of the merging processes were also found in the centre, like a disc produced from merger-triggered gas inflow or an infalling disc- or cluster-like central part of one of the merging components. A nuclear disc could explain the relatively thin ($\lesssim 8$ pc) and apparently elongated structure of the σ-drop. With a width of only $\lesssim 8$ pc this is probably the thinnest σ-drop ever detected. It is even narrower than the drop in NGC 1399 and also far less extended than the typical drops in spiral galaxies. As its size is comparable to the spatial resolution it is possibly unresolved. On the other hand rotation would be seen if there really was a disc.

The centre could still be unrelaxed and in the process of merging. This could induce signatures in the kinematics like those seen in Fornax A. Simulations of unequal-mass mergers with a SMBH in the primary galaxy and a significantly smaller secondary galaxy (without black hole) were performed by Holley-Bockelmann & Richstone (2000). Based on this it depends critically on the orbital decay trajectory of the secondary, and also somewhat on the black hole mass and on the mass ratio of the galaxies, whether the secondary galaxy is destroyed by the merger or not. The time-scale for a stellar cluster falling in on a purely radial orbit – also suggested by Gebhardt et al. (2007) as one possible explanation for the σ-drop in NGC 1399 – however, seems to be much shorter (of the order of about 10^7 yr) than the date of the last merger ($\sim 10^9$ yr), if the conditions in Fornax A are similar as in Holley-Bockelmann & Richstone (2000). For nonradial mergers the formation of a rotating stellar disc is possible. As we do not know any of the details of the merger history with desirable precision for Fornax A, and as the time-scales seem to vary depending on the initial parameters of the merger simulation, it is impossible to tell whether the centre is still unrelaxed or not.

The effect of dust

Fornax A contains a lot of dust features (see lower panel of Fig. 5.3) and this dust might alter the kinematics. On the other hand observations in the K-band should minimize the effects of dust and indeed already the uncorrected SOFI K_s-band image shows only very little amounts of dust compared to optical *HST* images. Some dust is located close to the centre, but not in the central

5.3. STELLAR KINEMATICS

$\sim 2''$. Shaya et al. (1996) find a low colour excess in the central pixels of an *HST* WFPC $V-I$ image and conclude that here the dust extinction is only very light. Consistently the SINFONI 25mas and 100mas images also do not show any signs of dust (cf. Fig. 5.1). The 250mas image shows hints of dust in the SE region, spatially coincident with the high-σ region just outside the 100mas field of view. Therefore the dust outside the central $2-3''$ could be the reason for the larger average σ of the 250mas data, in agreement with the findings of Silge & Gebhardt (2003).

Due to the lack of dust in the 25mas image, dust does not seem to be a plausible explanation for the central σ structure. However, the scale of the effect of dust on the kinematics might depend on the location of the dust along the line of sight. E.g. a certain amount of dust directly behind the centre could have a noticeable effect without being noticed in the photometry at the given S/N. To further substantiate this, we tested the effects on the kinematics when dust obscures light from stars behind or in front of the centre of a galaxy using an N-body model realization of the best-fitting dynamical model (cf. Section 5.5.2) as described in Thomas et al. (2007). We calculated the kinematics of the model by obscuring the central part at different radii and with different opacities and found that only a noticeable decrease in σ can be induced by putting a very thick layer of dust in front of the centre, but in this case the surface brightness would be reduced by more than a magnitude. Therefore we can exclude that the structure of the 25mas kinematic fields is produced by dust.

Stellar populations

Another explanation might be that the stellar population is different in the central region. Blue stars in the nucleus could in principle also be responsible for the low colour excess Shaya et al. (1996) find. The K-band absorption line equivalent widths for Fornax A are presented in Section 5.4 in detail. The 100mas maps of the absorption line indices (Fig. 5.8) clearly show that the line indices change within the central $\sim 0.4''$. All indices show a decrease in this region, CO(2 − 0) as the deepest absorption feature shows it most clearly. Unfortunately this trend is not as clear in the 25mas indices, as here the S/N is only about half as in the 100mas data and the scatter therefore is much larger.

Thus a dynamically colder subsystem with accordingly different line indices could be an explanation of the distorted kinematics, but the S/N of the 25mas data unfortunately is not large enough to resolve any spatial coincidence with the v peak or σ drop. Similar conclusions have been made by Lyubenova et al. (2008) for NGC 1399, where the central σ-drop coincides with a drop of the Na I and CO indices.

If a changing stellar population in the central region has an effect on the kinematics or not was tested with the N-body model mentioned in the previous section. Making the stars in the central 0.1″ ten times brighter (corresponding to a ten times lower mass-to-light ratio) results only in very small changes of $\Delta\sigma = 4$ km s^{-1} and $\Delta h_4 = -0.002$, both within the errors of the data and much smaller than the observed variations. 100 times brighter stars would produce noticeable changes of $\Delta\sigma = 25$ km s^{-1} and $\Delta h_4 = -0.015$. Making the stars darker does not have an effect at all on the kinematics. As there is no photometric evidence of a very bright star cluster in the centre, we conclude that a stellar population alone, which is different from the surrounding population, is not the reason for the disturbed central kinematics.

AGN emission

A decrease of the line indices would also be expected for AGN, where the absorption features are diluted by the non-stellar emission of the active nucleus. In this case also the measurement of v and σ could be affected. As shown in Davies et al. (2007) the PSF can be reconstructed from the decrease of the CO equivalent width at the position of the point source. Comparing the PSF measured this way with the measured or the noisy reconstructed PSFs from Fig. 5.2 is difficult, because (1) we would need to measure the CO index of the unbinned spectra, which have a lower S/N than the binned ones and thus errors larger than the small ~ 1.0 Å decrease observed in the binned spectra, and (2) for the higher S/N 100mas exposure we do not exactly know the true PSF. From the binned spectra we cannot deduce the two-dimensional PSF. A rough estimate of the FWHM of the CO dip can be obtained from fitting the averaged CO indices of a ring (red points in Fig. 5.8). The resulting FWHM of 0.17″ in the 100mas scale is, considering the large errors introduced by the binning, well in agreement with the measured PSF. This would favour the AGN scenario.

It was found by Iyomoto et al. (1998) that the AGN in Fornax A, which powered the very luminous radio lobes, became dormant during the last 0.1 Gyr. Still the nucleus shows weak radio (Geldzahler & Fomalont, 1984) and weak X-ray emission with a spectrum typical for a low-luminosity AGN (Kim & Fabbiano, 2003) and is bright in the UV (Fabbiano et al., 1994). In high-S/N optical spectra several emission lines are present ([OIII], Hβ, NI), but very weak (Beuing et al., 2002). In the SINFONI spectra there are no apparent emission lines typical for an AGN present at the given S/N. Only weak molecular hydrogen ($1-0S(1)$ H$_2$ at $\lambda = 2.122$ μm) with a S/N around 5 is found in a region north-east of the nucleus (see Fig. 5.9), but does not seem to be associated with the central source. The total H$_2$ flux in the 100mas field of view is 5.65×10^{-18} W m^{-2} and in a 3″ aperture centred on the continuum peak it is 3.64×10^{-18} W m^{-2}. The H$_2$ could be

5.3. STELLAR KINEMATICS

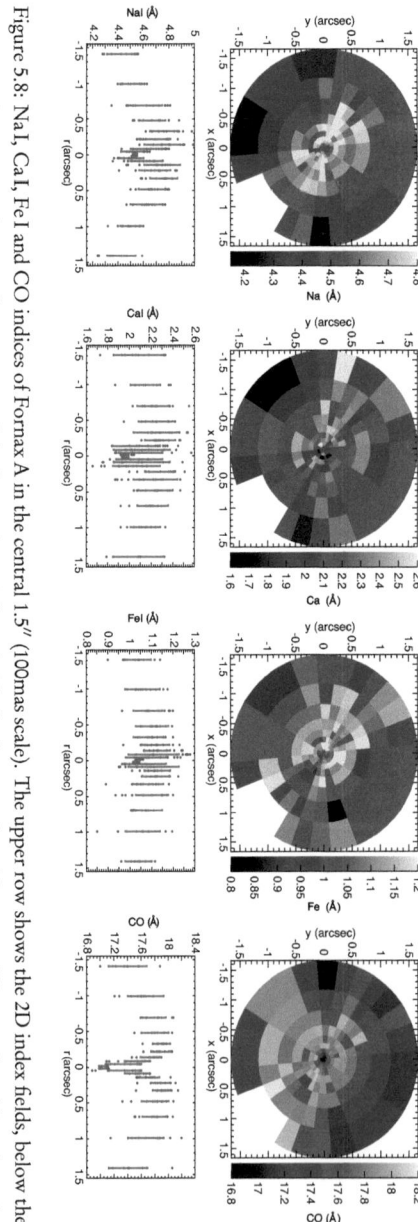

Figure 5.8: NaI, CaI, FeI and CO indices of Fornax A in the central 1.5″ (100mas scale). The upper row shows the 2D index fields, below the position-i diagrams are shown (where i is one of the four indices). The individual values are plotted in grey, while overplotted in red are the mean indices and their RMS of the bins belonging to a ring of radius r.

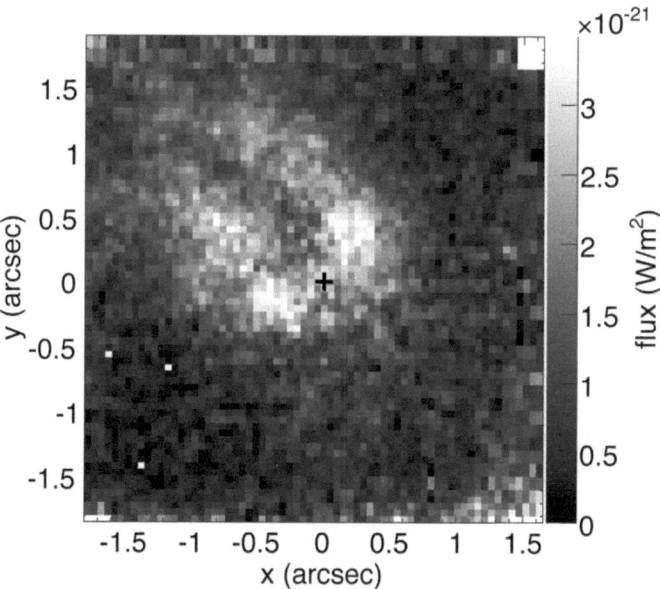

Figure 5.9: H$_2$ emission of Fornax A (100mas scale). The photometric centre is marked by a cross.

excited by X-rays, as a similar northeast-southwest elongation is present in Chandra data. This X-ray emission was associated with hot interstellar medium with a temperature of $kT = 0.62$ keV by Kim & Fabbiano (2003).

A weak AGN present in Fornax A would also become noticeable by a slight change of the continuum slope, making the nucleus redder. This is due to the AGN UV-continuum which heats surrounding dust that emits in the near-IR (Oliva et al., 1999). We indeed find that the slope of the continuum changes slightly in the inner $\approx 0.1''$ by no more than $2-3\%$.

Some AGN show a certain coronal emission line in the region of the CO bandheads, which alters the measured kinematics: [Ca VIII] at 2.3213 μm (Davies et al., 2006; Portilla et al., 2008). As it is located directly at the left edge of the second CO bandhead, already a very small contribution can have a significant effect on the measured kinematics. We tested this using two different approaches. First we tried to measure the kinematics using only the first bandhead. The velocity map in this case has a much larger scatter and shows many peaks of 20-30 km s^{-1} in the field of view. σ remains largely unchanged. When the first together with the third and fourth bandhead are used, the velocity map is smooth and shows no increase in the centre. The σ map is qualitatively similar to the one measured with the first two bandheads, but the absolute value of σ is much larger. The third and fourth bandheads are significantly distorted by sky emission lines and therefore do not give reliable results. The second approach was to use a standard star, convolve it with a LOSVD ($v = 0$ km s^{-1}, $\sigma = 230$ km s^{-1}, $h_3 = h_4 = 0$) and add the [Ca VIII] line with a typical FWHM of 150 km s^{-1}. The results are shown in Figs. 5.10 and 5.11. A [Ca VIII] contribution of $\lesssim 4\%$ would not be seen in the spectra or the fit at the given S/N, but can increase v by $\lesssim 25 - 30$ km s^{-1} and decrease σ by $\lesssim 10$ km s^{-1} (σ measured with zero [Ca VIII] contribution is already a few km s^{-1} lower than the real σ, see e.g. Fig. 3.5 in Section 3.4.2). Therefore a weak AGN with some [Ca VIII] contribution is the most logical explanation for the observed velocity increase. The observed σ decrease probably has for the most part a different origin (most probably a stellar population effect), as it is significantly larger than the $\lesssim 10$ km s^{-1} estimated from the simulations. In addition the high-v region only partly coincides with the low-σ region. The hint of a [Ca VIII] line signature appears when stacking all spectra within the region covered by the innermost bins and comparing it to a combination of all spectra within a ring further out ($0.11 < r < 0.17''$; see Fig. 5.12).

As accurate kinematics for the centre cannot be obtained from only one bandhead or by including the third and/or fourth bandhead, we will exclude the central bins in some of the dynamical models. As the black hole sphere of influence is large, this solution will not degrade the reliability of the resulting black hole mass significantly.

5.4 Line indices

The optical line strengths of Fornax A have been analysed extensively by Kuntschner (1998, 2000). They found a young (around 2 Gyr) and very metal-rich stellar population. The optical index gradients slightly decrease with radius. Unfortunately their data have a resolution of only $1''$, therefore they do not resolve the structure in the central $0.4''$. The age of a number of globular clusters in Fornax A was found to be about 3 Gyr (Goudfrooij et al., 2001). Thus a plausible scenario is that the globular clusters formed during the last major merger event 3 Gyr ago (Goudfrooij et al., 2001) and that the young stellar population formed from infalling molecular gas also as a result of this merger (Horellou et al., 2001).

The spectral features in the K-band can in principle be likewise interpreted in terms of age and metallicity. Unfortunately there is not such a sophisticated theoretical spectral synthesis model available yet as for optical indices, and studies of K-band indices are rare and often include only

CHAPTER 5. THE SUPERMASSIVE BLACK HOLE OF FORNAX A

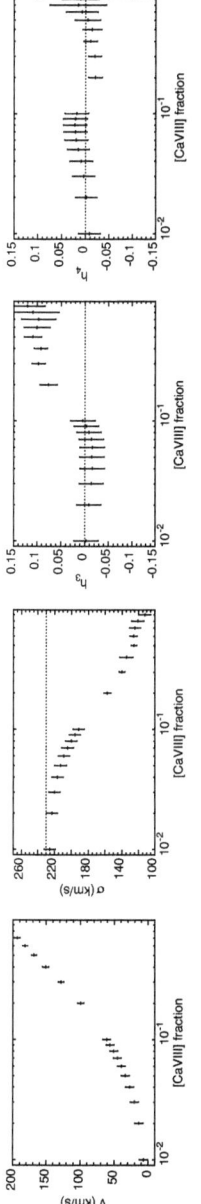

Figure 5.10: Mean v, σ, h_3 and h_4 as a function of [Ca VIII] contribution. The error bars are the 67% intervals obtained from fitting 100 realizations of the broadened template spectrum with added noise such that the S/N is similar to the central Fornax A spectra.

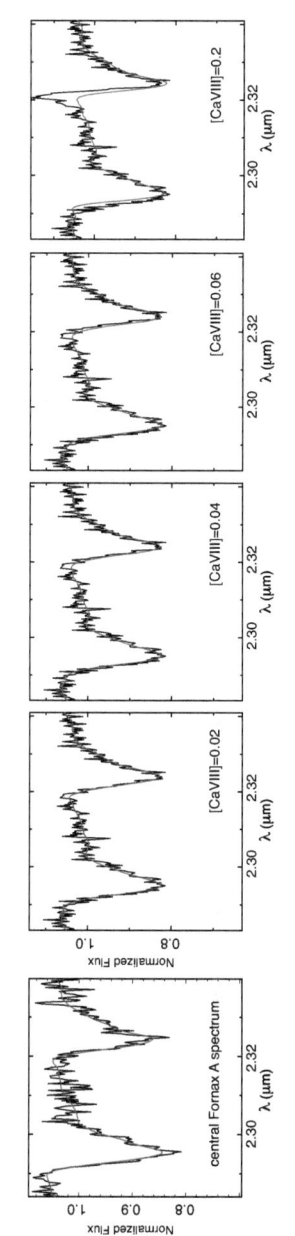

Figure 5.11: Fit (red) to broadened template star spectra (black) with different [Ca VIII] contributions compared to the fit to the central 25mas Fornax A spectrum with the largest measured velocity (leftmost plot).

5.4. LINE INDICES

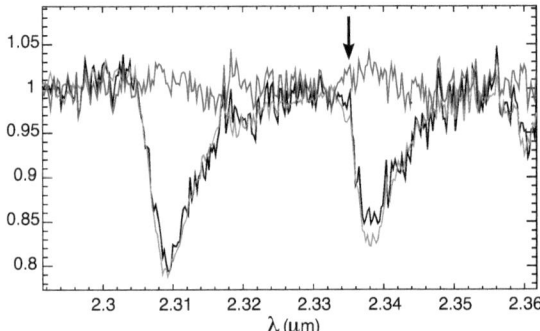

Figure 5.12: Combined spectra of the central region (black) at radii $r < 0.03''$ and of a ring at larger radii $0.11 < r < 0.17''$ (green). In red the residuals between the two spectra are plotted, shifted by +1. The position where the [Ca VIII] line is assumed to be is marked with an arrow.

small samples. A recent study of Silva et al. (2008) investigates the behaviour of the K-band indices of cluster stars of known age and metallicity and a sample of Fornax galaxies. Their published line indices of Fornax A are derived by extracting the mean spectrum over an aperture of $2 \times 3''$ from our SINFONI 100mas data.

In Fig. 5.8 we show the two-dimensional K-band line index maps (Na I, Ca I, Fe I and CO) of Fornax A for the 100mas scale, because the S/N is largest for this scale. The index definitions were taken from table 6 in Silva et al. (2008). Instead of Fe I1 and Fe I2 we use Fe I = (Fe I1 + Fe I2)/2, as they are highly correlated. The definition of the CO index is different from the definition of the CO equivalent width used above (Silge & Gebhardt, 2003). Here the continuum is defined by four blue continuum bands instead of a blue and a red one. The continuum on the red side is diminished by the CO absorption bands, therefore an accurate measurement of the CO index is possible by using blue continua only. Using the definition of Silge & Gebhardt (2003) nevertheless is sufficient to decide quickly which templates to use and to notice any particularities (see Table 5.1 for a comparison between the CO EWs of the kinematic template stars derived by the two different index definitions). As in Silva et al. (2008) we broadened the SINFONI spectra to match the ISAAC resolution of their data and to be able to directly compare our results to their findings.

The measured indices, averaged over the entire field of view, are in good agreement with the measurements of Silva et al. (2008), as shown in Tab. 5.2. In two dimensions, all indices show the same behaviour: a more or less clear dip in the centre and further out a negative gradient. This is most obvious in CO and Na I. Negative gradients are also present in the optical indices and it is

Table 5.2: Mean near-IR line indices of Fornax A.

	this work	Silva et al. (2008)
Na I	4.61 (0.15)	4.70 (0.14)
Ca I	2.14 (0.21)	2.48 (0.14)
Fe I1	1.44 (0.11)	1.491 (0.066)
Fe I2	0.73 (0.09)	0.843 (0.065)
CO	17.61 (0.33)	17.53 (0.30)

Note. The indices are given in units of Å. The middle column gives the near-IR line indices determined by averaging the values shown in Fig. 5.8 (upper row) and the according RMS errors. The indices determined by Silva et al. (2008) are given in the right column with their errors.

therefore reasonable to assume that the K-band index gradients can be likewise interpreted as age or metallicity gradients. Kuntschner (1998) find for the Fornax ellipticals that the index gradients are linked to a metallicity gradient, and not an age gradient. They assume that the negative gradients in Fornax A are likewise due to a metallicity gradient only, but their data do not go very far out in radius and therefore they cannot draw secure conclusions. Mármol-Queraltó et al. (2008) quantify a metallicity dependence of the CO index, which supports this interpretation.

What causes the dip in the very centre is less obvious. It might be one of the scenarios discussed above – AGN emission, dust or a different stellar population like a nuclear cluster or a population of younger, bluer stars of a post-starburst population. Fabbiano et al. (1994) detected a UV-bright unresolved central source with $r < 3$ pc, which according to them could be explained e.g. by a cluster of O stars or an AGN. The FWHM of the CO dip is very similar to the spatial resolution and thus unresolved, corresponding to an upper limit in size of $r \lesssim 4$ pc. Thus it is likely that the UV emission and the CO dip are caused by the same structure. Both early-type stars – due to the lack of typical late-type star absorption lines – and non-stellar emission would cause the observed decrease of all indices. Br γ absorption is the only strong feature in early-type K-band stellar spectra, Br γ emission is observed in many galaxies with nuclear activity or star formation, thus this feature could be used to distinguish between these two possibilities. In the central SINFONI spectra we do not detect Br γ neither in emission nor in absorption. Thus emission or absorption is either not present, too weak to be detected, or Br γ emission and absorption just compensate. [Ca VIII] emission caused by the AGN is also very well hidden (see Fig. 5.12 and discussion in Section 5.3.4). Due to the broader PSF its influence is eliminated in the 100mas data. Other emission lines typically present in AGN spectra are absent as well in the central region. Weak H_2 emission is present in the north-eastern region, but not correlated to the central source (see Fig. 5.9). In order to resolve the reasons that cause the central line index dip and kinematics, very high

5.5 Dynamical models

Based on the $M_\bullet$-σ relation of Tremaine et al. (2002) and our σ measurements a black hole with a mass around 2×10^8 $M_\odot$ would be expected. Earlier models by Shaya et al. (1996) suggest a ten times higher black hole mass, but they rely on larger σ values. Davies (2000) observed Fornax A with the MPE 3D IFS (Weitzel et al., 1996) and measured v and σ using the CO bandheads at 2.2 µm. Their σ is lower than the values used by Shaya et al. (1996) and comparable to our values. Together with the models of Shaya et al. (1996) they obtain a SMBH mass of $\lesssim 10^9$ $M_\odot$. However, they cannot exclude that radial anisotropy may account for this large mass. A lower limit of the central mass concentration of 75 $M_\odot$ was estimated by Fabbiano et al. (1994) based on the measured nuclear UV emission.

We used the axisymmetric code of Gebhardt et al. (2000a, 2003) in the version of Thomas et al. (2004) to determine the mass of the SMBH in Fornax A. It is based on the Schwarzschild (1979) orbit superposition technique and comprises following steps: (1) Calculation of the gravitational potential of the galaxy from the stellar mass density ρ_* using a trial black hole mass $M_\bullet$ and mass-to-light ratio Υ. (2) Generation of an orbit library for this potential and construction of a weighted superposition of orbits that best matches the observational constraints. (3) Repetition of the first two steps with different values for Υ and $M_\bullet$ until the eligible parameter space is systematically sampled. The best-fitting parameters then follow from a χ^2-analysis. Our orbit libraries contain around 2×7000 orbits. The deprojected luminosity density is a boundary condition and hence exactly reproduced, while the LOSVDs are fitted in 25 velocity bins between -880 and $+880$ km s^{-1} with a bin width of ≈ 75 km s^{-1}. Special care was taken when implementing the PSF. Due to its asymmetric shape the PSF was not fitted, but the models were rather convolved with the two-dimensional image of the PSF reference star directly (cf. Chapter 4 or Nowak et al. 2007). All modelling was done with minimal regularisation.

A big advantage of integral-field data compared to longslit data only is the assurance of whether the assumption of axisymmetry is legitimate by comparing both the kinematics and the results of the dynamical modelling of all four quadrants. In case all quadrants produce the same results, the data can be folded and averaged, so that the errors are reduced. This results in a very large number of models. For Fornax A we calculated around 5000 models in total, covering a broad parameter space in $M_\bullet$ ($0 - 5 \times 10^8$ $M_\odot$) and Υ ($0.55 - 0.95$) for each quadrant, several combinations of data

CHAPTER 5. THE SUPERMASSIVE BLACK HOLE OF FORNAX A

sets and PSFs using an inclination of 90°. At the beginning we computed some test models using either the longslit data only or the SINFONI data only (Section 5.5.1) in order to constrain the Υ range. As a second step each of the four quadrants was modelled separately using the longslit data and either the 25mas data or the 100mas data or a combination of the two (Section 5.5.2). This step is essential as it reveals possible inconsistencies between the data sets, non-axisymmetries and other problems. The assumed diameter of the sphere of influence is around 0.46", resolved by both SINFONI data sets. The 250mas data were not included in the modelling, as they do neither resolve the sphere of influence nor do they add significantly more information in the outer parts that would justify the enhanced amount of computing time. After a careful examination of the results and verifying the consistency of the data with axisymmetry, the four quadrants were averaged and modelled for the same combination of data sets as for the single quadrants.

5.5.1 The stellar dynamical K_s-band mass-to-light ratio Υ

In order to check if the stellar mass-to-light ratios Υ obtained for individual datasets, agree and if they are consistent with Υ from population synthesis models (Maraston, 1998, 2005), we determine Υ by modelling single data sets.

Υ from longslit data

First a number of models were calculated using exclusively the longslit data without black hole covering a broad range in Υ (in units of $M_\odot/L_\odot^{K_s}$). The longslit data were truncated at different radii r_{trunc} between 14" and 45". The resulting best-fitting Υ (Fig. 5.13) is roughly constant with r_{trunc} and $\approx 0.75 - 0.8$ for longslit data at $r_{trunc} \lesssim 31" \approx 2.8$ kpc, which roughly corresponds to the effective radius of Fornax A ($R_e = 36^{+5.7}_{-11.2}$ arcseconds, Bedregal et al. 2006). At larger truncation radii Υ increases strongly, a sign that here the dark halo starts to have an effect. This is consistent with the results of Thomas et al. (2007), who found dark matter fractions of $\approx 30\%$ at 3 kpc in similar bright Coma galaxies. Thus in the following only longslit data at $r < 31"$ will be used for the dynamical modelling.

Υ from SINFONI data

For the SINFONI 100mas data alone the best-fitting Υ is slightly lower (≈ 0.7 in units of $M_\odot/L_\odot^{K_s}$, see Fig. 5.14), but within the errors consistent with Υ derived from the longslit data alone. The contours are strongly elongated towards large Υ with slightly decreasing $M_\bullet$, as large Υ in the centre

5.5. DYNAMICAL MODELS

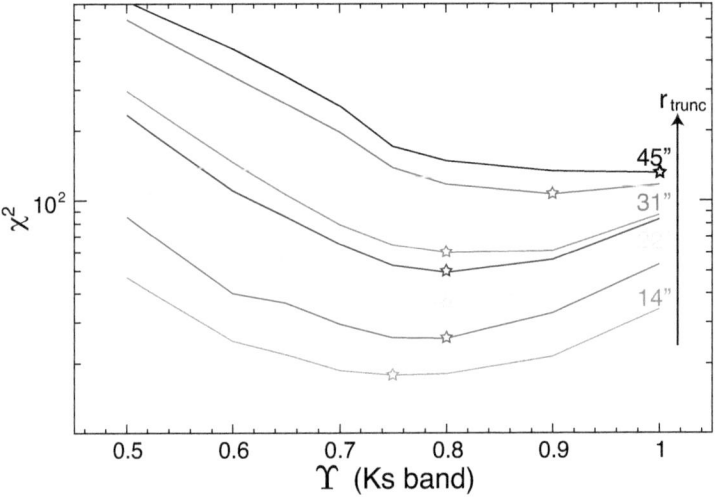

Figure 5.13: χ^2 as a function of mass-to-light ratio Υ for models calculated with the longslit data alone, truncated at different radii r_{trunc} and without black hole. The models with the smallest χ^2 are marked with a star.

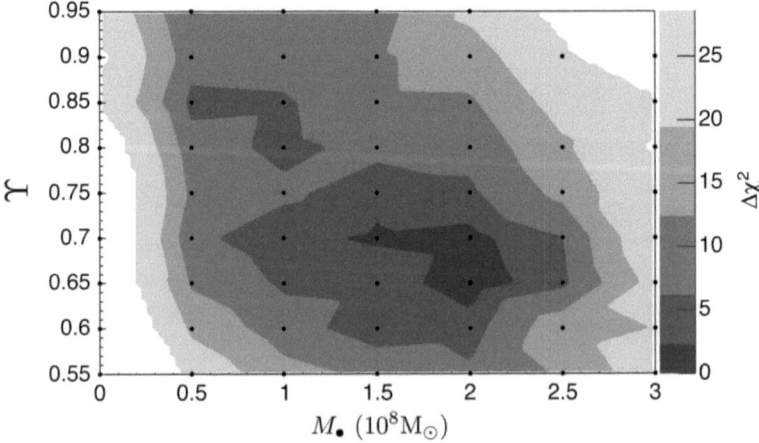

Figure 5.14: $\Delta\chi_0^2 = \chi^2 - \chi_{\min}^2$ as a function of Υ (K_s-band) for models calculated with the SINFONI 100mas data alone and a small number of black hole masses $M_\bullet$. The coloured regions are the $1-5\,\sigma$ errors in $\Delta\chi_0^2$. Each black dot represents a model.

can somewhat compensate a lower $M_\bullet$. This degeneracy between $M_\bullet$ and Υ is even stronger for the tiny field of view 25mas data alone, where Υ cannot be constrained very well. This emphasizes the importance of including the longslit data in the dynamical modelling, which makes the degeneracy virtually disappear.

In order to be able to compare our stellar dynamical K_s-band Υ to predictions from population synthesis models, we need to scale it to units of $M_\odot/L_\odot^K$ by multiplying it with the ratio of the bandwidths:

$$\Upsilon_K = \Upsilon \frac{\Delta K_s}{\Delta K} = \Upsilon \frac{0.275\,\mu m}{0.390\,\mu m} \approx 0.5 \qquad (5.4)$$

For a 2 − 3 Gyr old population with a high metallicity of $[Z/H] = 0.3 - 0.45$ (Kuntschner, 1998, 2000) the models of Maraston (1998, 2005) predict $\Upsilon_K \approx 0.3 - 0.6$ (Salpeter IMF) and $\Upsilon_K \approx 0.2 - 0.4$ (Kroupa IMF). Thus the measured stellar dynamical Υ is in good agreement with a Salpeter IMF but larger than predicted for a Kroupa IMF. When taking into account the error in the measured dynamical Υ and a possible error in the distance assumed for Fornax A, then the dynamical Υ is marginally in agreement also with the Kroupa IMF. Cappellari et al. (2006) also find that the dynamical Υ for E/S0 galaxies are in agreement or larger than the Υ predicted using a Kroupa IMF.

5.5.2 The black hole mass $M_\bullet$

$M_\bullet$ from the 25mas and the longslit data

After having constrained the possible Υ range we now focus on the determination of the mass of the central black hole. We modelled the 25mas data set together with the longslit data for each quadrant separately and at an inclination of 90°. The resulting values for $M_\bullet$ and Υ with the 3 σ error bars are listed in Tab. 5.3. The best-fitting black hole masses and mass-to-light ratios are in the range $M_\bullet = (2.0 - 3.25) \times 10^8\,M_\odot$ and $\Upsilon = 0.675 - 0.775$ and agree very well within 2 σ ($\Delta\chi_0^2 = \chi^2 - \chi^2_{min} = 6.2$) errors.

$M_\bullet$ from the 100mas and the longslit data

The same was done using only the 100mas data set together with the longslit data and the measured PSF. Both $M_\bullet$ and Υ are in agreement between the different quadrants, but slightly smaller than for the 25mas data ($M_\bullet = 1.0 - 2.0 \times 10^8\,M_\odot$, $\Upsilon = 0.65 - 0.7$). Within 2 − 3 σ errors they still agree with the 25mas results. Note that the sphere of influence is well resolved in both cases.

5.5. DYNAMICAL MODELS

To figure out whether the uncertainties in the PSF could raise a bias in $M_\bullet$, the same models were calculated with the 25mas PSF convolved with the kernel from Fig. 5.2c. The results are virtually identical (see Tab. 5.3). Therefore, slight uncertainties in the PSF shape do not have a noticeable effect on $M_\bullet$ or Υ.

As a cross-check, the 25mas data, binned 4×4 to achieve the 100mas spaxel size and convolved with the kernel, were also modelled using the convolved 25mas PSF. In this case the confidence intervals are much wider and the best-fitting values agree within $1 - 2\, \sigma$ with both the 25mas and the 100mas results.

The difference between the two scales could be related to the different spatial coverage, although in that case we would expect a similar result with larger error bars for the 25mas models. An increase of Υ towards the centre could also be an explanation. The global Υ would basically not change in that case, and a larger black hole mass could mimic the Υ gradient. A triaxial structure could also cause a systematic difference between the 25mas and the 100mas scale.

Due to the good overall agreement of the different quadrants this implies that the assumption of axisymmetry is justified and therefore the four quadrants of each data set were folded and averaged. The results of the averaged data agree with the results of the individual quadrants.

$M_\bullet$ from the longslit and a combination of the 25mas and the 100mas data

In order to take advantage of the high spatial resolution of the 25mas data set and to constrain the orbital distribution of the galaxy adequately the 25mas data were combined with the 100mas data and this combined data set was modelled together with the longslit data. The spatial region covered already by the 25mas data was not considered in the 100mas data set. The results of the dynamical modelling are shown in Tab. 5.3 and Fig. 5.15, where $\Delta\chi_0^2$ is plotted as a function of $M_\bullet$ and Υ with error contours for two degrees of freedom. All four quadrants deliver the same results within at most $3\, \sigma$ errors when all data bins are considered (the mean black hole mass is $\langle M_\bullet \rangle = 1.3 \times 10^8\, M_\odot$ with a corresponding $\mathrm{rms}(M_\bullet) = 0.4 \times 10^8\, M_\odot$ and the mean K_s-band mass-to-light ratio is $\langle \Upsilon \rangle = 0.68$ with a corresponding $\mathrm{rms}(\Upsilon) = 0.03$). When the inner two radial bins are excluded the results are almost identical ($\langle M_\bullet \rangle = 1.2 \times 10^8\, M_\odot$, $\mathrm{rms}(M_\bullet) = 0.5 \times 10^8\, M_\odot$, $\langle \Upsilon \rangle = 0.70$ and $\mathrm{rms}(\Upsilon) = 0.02$). Only the individual $\Delta\chi_0^2$ contours are wider because of the decreased resolution in the centre.

As the single quadrants agree well with each other it is legitimate to just model the LOSVDs folded and averaged over all quadrants. This is shown in Fig. 5.16 for both the entire data set (Fig. 5.16a) and with the central bins excluded (Fig. 5.16b). After the averaging the central velocity peak

CHAPTER 5. THE SUPERMASSIVE BLACK HOLE OF FORNAX A

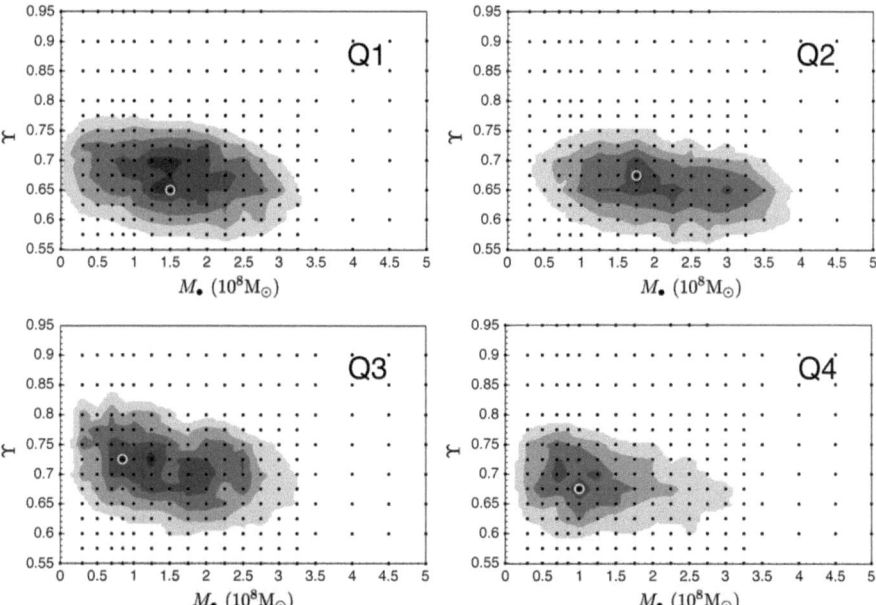

Figure 5.15: Models calculated for all four quadrants using a combination of longslit data out to $31''$, 100mas and 25mas SINFONI data. For each quadrant $\Delta\chi_0^2 = \chi^2 - \chi_{\min}^2$ is plotted as a function of the black hole mass $M_\bullet$ and the K_s-band mass-to-light ratio Υ. The coloured regions are the $1-5\,\sigma$ confidence intervals for two degrees of freedom ($\Delta\chi_0^2 = 2.28, 6.20, 12.43, 19.44$ and 28.65). Each model that was calculated is marked as a black dot, the best-fitting model is encircled by a white ring. The $\Delta\chi_0^2$ contours are unsmoothed, which sometimes results in disconnected $1\,\sigma$ regions due to noise in the models (e.g. in Q3).

5.5. DYNAMICAL MODELS

disappears, as the LOSVDs of the third and fourth quadrant are mirrored. The results therefore are in both cases very similar. They also agree very well with the single quadrants. The black hole mass of Fornax A, derived from modelling the averaged LOSVDs of the combined 25mas and 100mas SINFONI data set and longslit data, is $M_\bullet = 1.50^{+0.75}_{-0.80} \times 10^8$ $M_\odot$ and the according mass-to-light ratio is $\Upsilon = 0.650^{+0.075}_{-0.050}$ (3 σ errors). The fit of the best-fitting model to v, σ, h_3 and h_4 along the major and the minor axis is shown in Fig. 5.17. Note that the best fit without black hole would be hardly distinguishable from the shown fit with black hole (see discussion below and Fig. 5.18). The result does not change when the inner two radial bins, where the AGN emission distorts the CO absorption bands are excluded. Only the error bars become somewhat larger ($M_\bullet = 1.50^{+1.25}_{-1.20} \times 10^8$ $M_\odot$ and $\Upsilon = 0.725^{+0.025}_{-0.125}$). Note that the 1 σ errors derived from a smoothed version of the χ^2 profile agree well with the RMS of the four quadrant's solution (see above).

In order to illustrate the significance of the result and where the influence of the black hole is largest, Fig. 5.18 shows the χ^2 differences between the best-fitting model without black hole and the best-fitting model with black hole for all LOSVDs of the averaged quadrant, analogous to Fig. 4.7 in Chapter 4. As in the case of NGC 4486a the largest black hole signature is found within about 2 spheres of influence and in particular along the major axis. For about 85% of all bins the model with black hole produces a fit to the LOSVD better than the model without black hole. Adding a black hole improves the fit everywhere and not only in the centre, because orbit-based models have a lot of freedom and will choose a different orbit distribution if no black hole is assumed. Or put in another way: a model without a black hole will not just be worse in the very centre but will be worse over a relatively large area of the galaxy. For the bins with the largest $\Delta\chi^2$ along the major axis the LOSVD and the fits with and without black hole are shown in the left part of Fig. 5.18 together with the corresponding $\Delta\chi_i^2$ as a function of line-of-sight velocity. The largest χ^2 differences appear in the high-velocity tails of the LOSVDs. The total χ^2 difference, summed over all LOSVDs, between the best model with black hole and the best model without black hole is $\Delta\chi^2 = 54.7$, which corresponds to about 7.1 σ. The total χ^2 values for the models are around 450. Together with the number of observables (60 LOSVD bins ×25 velocity bins) this gives a reduced χ^2 of ≈ 0.3. This is a reasonable value as the effective number of observables is smaller due to the smoothing (Gebhardt et al., 2000a).

For the best-fitting model ($\Upsilon = 0.65$) the total stellar mass within 1 sphere of influence, where the imprint of the black hole is strongest, is $M_* = 1.11 \times 10^8$ $M_\odot$. If the additional mass of $M_\bullet = 1.5 \times 10^8$ $M_\odot$ was entirely composed of stars, the mass-to-light ratio would increase to $\Upsilon = 1.53$ (corresponding to $\Upsilon_K = 1.08$). This value would be typical for an old stellar population (around 8 Gyr for a Salpeter IMF, or 13 Gyr for a Kroupa IMF), which has not been found in Fornax A (Goudfrooij et al., 2001; Kuntschner, 2000).

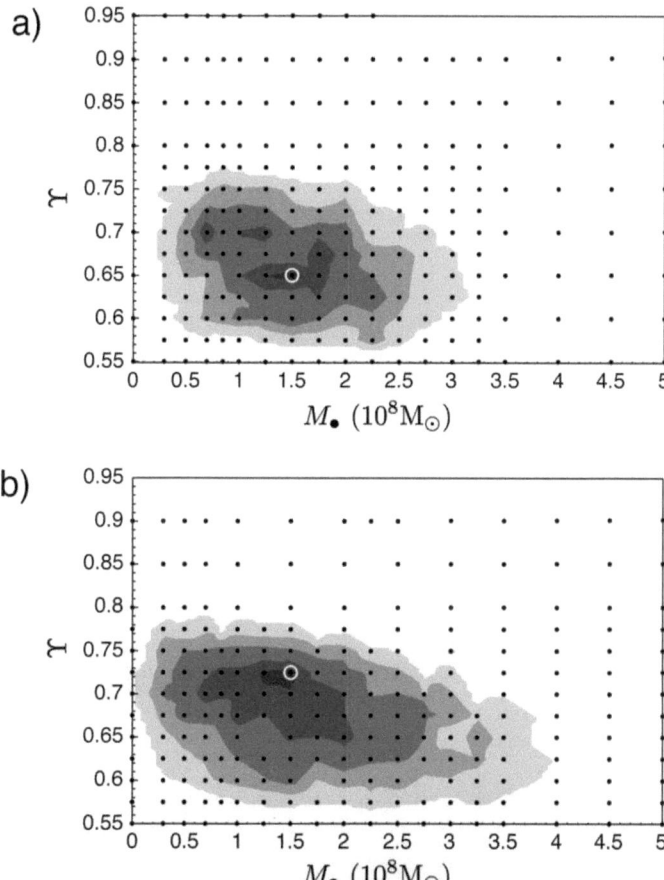

Figure 5.16: Same as Fig. 5.15 (longslit, 100mas and 25mas data) but for the averaged quadrant with (a) or without (b) the bins of the central two radii.

5.5. DYNAMICAL MODELS

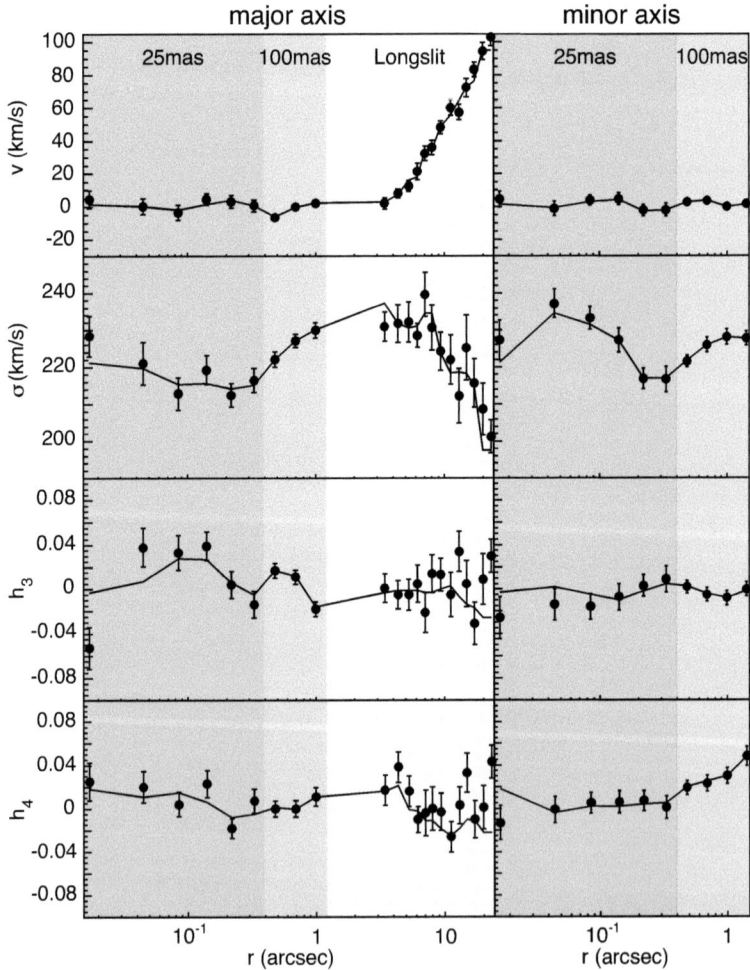

Figure 5.17: Fit (solid line) of the best model with $M_\bullet = 1.5 \times 10^8$ $M_\odot$ and $\Upsilon = 0.65$ to the major and minor axis kinematics (points).

Table 5.3: Resulting black hole masses $M_\bullet$ and K_s-band mass-to-light ratios Υ with the according 3 σ ($\Delta\chi_0^2 = 12.43$) errors for all modelled data sets and quadrants.

Data set	Quadrant	$M_\bullet$ [10^8 M$_\odot$]	$M_{\bullet,-}^a$	$M_{\bullet,+}^b$	Υ	Υ_-^c	Υ_+^d
25mas[e] +LS[f]	1	2.0	0.3	3.5	0.725	0.625	0.775
	2	3.25	1.5	4.0	0.675	0.625	0.75
	3	2.5	0.7	3.25	0.75	0.675	0.825
	4	2.0	0.3	4.0	0.775	0.675	0.825
	1−4 a[g]	1.75	0.7	3.0	0.7	0.625	0.75
	1−4 a c[h]	3.0	0.3	4.5	0.7	0.65	0.75
100mas[i] +LS	1	1.5	0.7	2.5	0.65	0.625	0.7
	2	1.25	0.85	2.75	0.7	0.6	0.7
	3	1.0	0.3	2.0	0.675	0.65	0.75
	4	2.0	0.5	3.0	0.675	0.625	0.725
	1−4 a	1.0	0.3	1.75	0.65	0.6	0.725
100mas conv.[j] +LS	1	1.5	0.5	2.0	0.65	0.65	0.7
	2	1.5	0.7	2.5	0.7	0.65	0.7
	3	1.0	0.3	1.5	0.7	0.65	0.75
	4	2.0	0.5	2.5	0.7	0.65	0.7
25mas conv.[k] +LS	1	2.0	0.0	4.0	0.7	0.65	0.8
	2	1.5	0.3	4.0	0.7	0.65	0.8
	3	0.7	0.3	3.5	0.8	0.7	0.85
	4	1.0	0.3	2.5	0.75	0.7	0.8
25mas+100mas+LS	1	1.5	0.3	2.75	0.65	0.625	0.725
	2	1.75	1.0	3.5	0.675	0.625	0.7
	3	0.85	0.3	2.5	0.725	0.65	0.8
	4	1.0	0.5	1.75	0.675	0.625	0.75
	1−4 a	**1.5**	**0.7**	**2.25**	**0.65**	**0.6**	**0.725**
	1 c	1.5	0.3	3.0	0.7	0.625	0.75
	2 c	1.75	0.5	3.0	0.675	0.625	0.725
	3 c	0.85	0.3	2.5	0.725	0.675	0.8
	4 c	0.7	0.3	2.0	0.7	0.65	0.75
	1−4 a c	1.5	0.3	2.75	0.725	0.6	0.75

Note. When only one SINFONI data set was included in the modelling, the $M_\bullet$-Υ parameter space was sampled more coarsely than for the combined SINFONI data set, with only about half the number of models per quadrant. This can sometimes result in the lower or upper limits being identical to the best-fitting values.

[a] Lower 3 σ limit of $M_\bullet$ [10^8 M$_\odot$]; [b] Upper 3 σ limit of $M_\bullet$ [10^8 M$_\odot$]; [c] Lower 3 σ limit of Υ; [d] Upper 3 σ limit of Υ; [e] SINFONI data with a field of field of view of 0.8 × 0.8″; [f] Longslit data from the Siding Spring 2.3m telescope (Saglia et al., 2002); [g] The LOSVDs of all four quadrants were averaged and then modelled; [h] Central two radial bins excluded due to AGN contamination; [i] SINFONI data with a field of view of 3.0 × 3.0″ together with the measured PSF were modelled; [j] 100mas SINFONI data were modelled together with the PSF from Fig. 5.2d; [k] 25mas SINFONI data, binned to the 100mas spaxel scale. Data and PSF were then convolved with the kernel from Fig. 5.2c.

5.5. DYNAMICAL MODELS

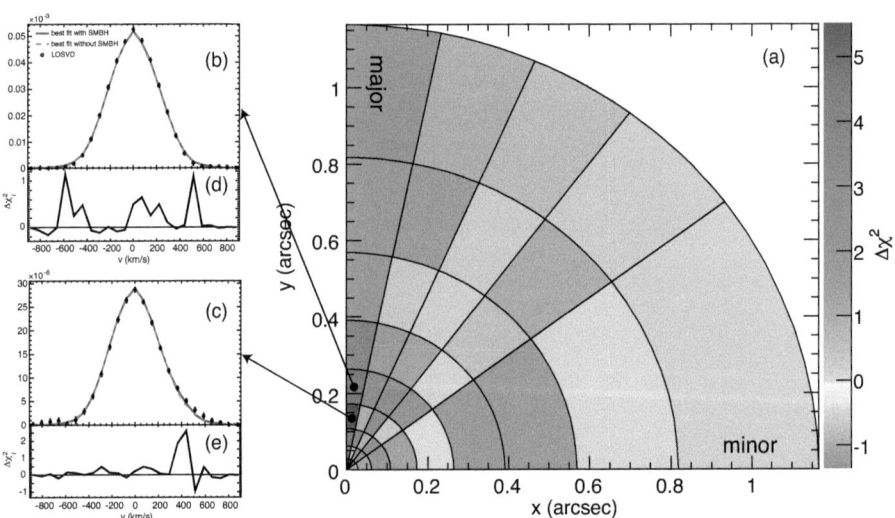

Figure 5.18: Panel (a): χ^2 difference between the best-fitting model without black hole and the best-fitting model with black hole ($\Delta\chi^2 = \sum_i \Delta\chi_i^2 = \sum_{i=1}^{25} \left(\chi_{i,\text{noBH}}^2 - \chi_{i,\text{BH}}^2\right)$ over all 25 velocity bins) for all LOSVDs of the averaged quadrant (longslit, 100mas and all 25mas SINFONI data). Bins where the model with black hole fits the LOSVD better are plotted in green the others in orange. Panels (b) and (c): For the radii with the largest positive χ^2 difference in panel (a) the LOSVD (open circles with error bars, normalised as in Gebhardt et al. (2000a)) and both fits (with black hole, full green line, and without black hole, dashed orange line) are shown with the corresponding $\Delta\chi_i^2$ plotted below (Panels (d) and (e)).

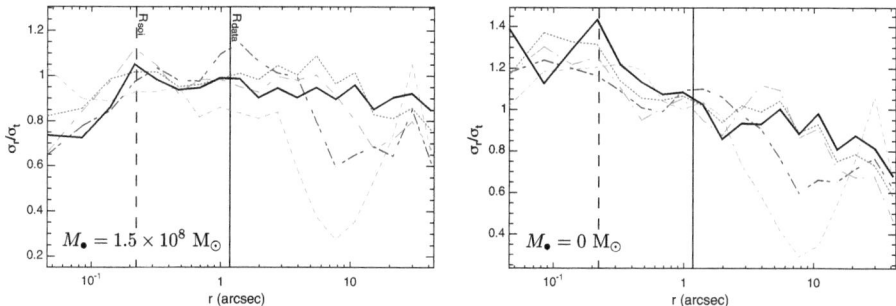

Figure 5.19: Radial over tangential anisotropy as a function of radius for the best-fitting model with black hole (left) and the best-fitting model without black hole (right). The values along the major axis are plotted in black, those along the minor axis in cyan and the other colours represent the position angles in between. The vertical dashed line indicates the radius of the sphere of influence and the vertical solid line marks the radius out to which the SINFONI data used in the models extend.

Fig. 5.19 shows the anisotropy profiles for the best-fitting model with black hole and the best-fitting model without black hole. The models with black hole become tangentially anisotropic in the centre ($r \lesssim 0.2''$), while the models without black hole show a certain degree of radial anisotropy ($\sigma_r/\sigma_t \approx 1.3$ where $\sigma_t = [(\sigma_\theta^2 + \sigma_\phi^2)/2]^{1/2}$). This behaviour is not surprising, as a central velocity dispersion increase can be modelled with either a large black hole mass or with radial orbits. However, the no black hole case is a significantly poorer fit than the best-fitting case with black hole. At large radii the tangential bias is decreased with an increasing black hole mass, but the anisotropy profile in this region is difficult to interpret since we did not include a dark halo.

5.6 Summary and discussion

We have obtained near-IR integral field data with three different spatial resolutions $\geqslant 0.085''$ for the merger remnant Fornax A. Stellar rotation is only detected in the outer parts ($r \gtrsim 1.5''$). The stellar velocity dispersion slightly decreases from large to smaller radii. In the highest resolution data it shows two peaks in the centre, separated by a narrow low-σ region. The average dispersions are $\sigma = 226 \pm 9$ km s^{-1} in an $8''$ diameter aperture, $\sigma = 221 \pm 4$ km s^{-1} in a $3.0''$ diameter aperture and $\sigma = 218 \pm 8$ km s^{-1} in an $0.8''$ diameter aperture. A low-luminosity AGN is likely to be present and distort the stellar kinematics in the central $\lesssim 0.06''$, where weak [Ca VIII] emission in the region of the second CO bandhead probably induces a velocity peak. The σ-drop can be

5.6. SUMMARY AND DISCUSSION

at most partially attributed to this emission line. Either the AGN continuum emission or a cold stellar subsystem or both could be the cause.

Near-IR line indices were measured using the high-S/N, intermediate resolution data to trace stellar populations. All indices show the same trend, a negative gradient and in the centre an unresolved depression. The negative gradient of the CO index can be interpreted as a metallicity gradient. The central drop (~ 1 Å in CO) could be, like the σ drop, caused by the low-luminosity AGN, a subsystem with a different stellar population, or a combination of the two.

Using an axisymmetric Schwarzschild code we find a black hole with the mass $M_\bullet = 1.5^{+0.75}_{-0.8} \times 10^8$ M$_\odot$ (3 σ C.L. for two degrees of freedom), when the 25mas, the 100mas and the longslit data are included. The same mass with slightly larger error bars is found when the innermost bins, where the kinematics is distorted by the AGN, are excluded from the dynamical modelling. The assumption of axisymmetry is justified as consistent results were obtained when modelling single quadrants. The mean black hole mass obtained from the four single quadrants is $\langle M_\bullet \rangle = 1.3 \times 10^8$ M$_\odot$ with rms($M_\bullet$) = 0.4×10^8 M$_\odot$ and the mean K_s-band mass-to-light ratio is $\langle \Upsilon \rangle = 0.68$ with rms(Υ) = 0.03, in agreement with the 1 σ limit derived from the analysis of the smoothed χ^2-profile for the averaged data. The PSF of the 100mas data could not be measured very accurately, but this does not seem to have noticeable effects. Modelling the 100mas data with the measured or a reconstructed, noisier PSF (both have about the same FWHM) does not yield different results.

We find a dynamical K_s-band mass-to-light ratio of $\Upsilon = 0.65^{+0.075}_{-0.05}$. When the inner two radial bins are not included it is somewhat larger ($\Upsilon = 0.725^{+0.025}_{-0.125}$). This is in agreement with stellar population models assuming a Salpeter IMF, but only marginally consistent with a Kroupa IMF. A dark halo plays a significant role only outside $\sim 31''$, close to the effective radius ($R_e = 36''$, Bedregal et al. 2006). We therefore included neither a dark halo nor longslit data at radii larger than $31''$.

The black hole mass expected from the $M_\bullet$-σ relation (Tremaine et al., 2002) is $(2.2 \pm 0.4) \times 10^8$ M$_\odot$. Note that estimating σ_e, the luminosity-weighted velocity dispersion within R_e, from Fig. 5.17 gives the same value as σ measured in the $8''$ aperture quoted above. This mass is consistent with our measurements within the errors. Fornax A is only one of two galaxies that underwent a major merger a few Gyr ago where a black hole mass has been measured (the other one is Cen A, e.g. Neumayer et al. 2007). Both seem to be consistent with the $M_\bullet$-σ relation, which holds important implications for the growth of the black hole and its surrounding bulge. Despite still showing obvious characteristics of the merging process – like shells and ripples in the outer envelope and disordered dust features throughout the entire galaxy – the black hole of Fornax A has about the mass expected for a normal elliptical. This suggests that bulges and black holes grow approximately synchronously in the course of a merger, or that the growth is regulated during the active phase.

The black hole mass of Fornax A is, however, not in agreement with the relations between black hole mass and K-band luminosity L_K or bulge mass M_{bulge} of Marconi & Hunt (2003). The total 2MASS K_s-band magnitude $m_{K_s} = 5.587$ corresponds to $\log(M_{bulge}/M_\odot) \approx 11.4$ for our $\Upsilon_{K_s} = 0.65$. For this high bulge mass a black hole mass of $\sim 6 \times 10^8$ $M_\odot$ would be expected, a factor ~ 4 larger than what we measured. Interestingly the situation is similar for Cen A. With the K-band magnitude given in Marconi & Hunt (2003) and $\Upsilon_K = 0.65$ (Silge et al., 2005) the bulge mass is $\log(M_{bulge}/M_\odot) \approx 10.9$ and a black hole with $M_\bullet \approx 1.9 \times 10^8$ $M_\odot$ would thus be expected, a factor ~ 3.5 larger than the mass of 5.5×10^7 $M_\odot$ obtained by Cappellari et al. (2009). Using L_K instead of the bulge mass M_{bulge}, Fornax A and Cen would be placed a factor of 7 respectively 5 below the $M_\bullet$-L_K relation of Marconi & Hunt (2003). In this context it would be interesting to reconsider the relationship between $M_\bullet/M_{bulge}$ and galaxy age found by Merrifield et al. (2000). Fornax A and Cen A both have a very small $M_\bullet/M_{bulge}$ and they would be among the youngest galaxies in their sample. Although the number of merger galaxies with measured black hole masses is fairly small, a correlation between the time of the last major merger and $M_\bullet$ might start to emerge.

In order to draw more stringent conclusions on the connections between merging, bulge growth, black hole growth and nuclear activity more dynamical black hole mass measurements in merger remnants and in luminous AGN are necessary. We were able to show that reliable black hole masses can be derived via stellar dynamical modelling even if nuclear activity makes dynamical measurements in the centre impossible, as long as the spatial resolution is high enough that the AGN signature is well within the sphere of influence.

Acknowledgments

We would like to thank the Paranal Observatory Team for support during the observations. We are very grateful to Harald Kuntschner for providing us the code to measure near-IR line indices, to Mariya Lyubenova for helping us implementing the software and numerous discussions about line indices, and to Yuri Beletsky for providing us the SOFI images of Fornax A. Furthermore we thank Peter Erwin and Robert Wagner for stimulating discussions. Finally we thank the referee for his useful comments. This work was supported by the Cluster of Excellence: "Origin and Structure of the Universe" and by the Priority Programme 1177 of the Deutsche Forschungsgemeinschaft.

6

Do black hole masses scale with classical bulge luminosities only? The case of the two composite pseudobulge galaxies NGC 3368 and NGC 3489[1]

Abstract

It is now well established that all galaxies with a massive bulge component harbour a central supermassive black hole (SMBH). The mass of the SMBH correlates with bulge properties such as the bulge mass and the velocity dispersion, which implies that the bulge and the central black hole of a galaxy have grown together during the formation process. As part of an investigation of the dependence of the SMBH mass on bulge types and formation mechanisms, we present measurements of SMBH masses in two pseudobulge galaxies. The spiral galaxy NGC 3368 is double-barred and hosts a large pseudobulge with a tiny classical bulge component at the very centre. The S0 galaxy NGC 3489 has only a weak large-scale bar, a small pseudobulge and

---

[1] A revised version of this chapter has been published as: "Nowak N., Thomas J., Erwin P., Saglia R. P., Bender R., Davies R. I., 2010, MNRAS, 403, 646"

6.1. INTRODUCTION

a small classical bulge. Both galaxies show weak nuclear activity in the optical, indicative of the presence of a supermassive black hole. We present high resolution, adaptive-optics-assisted, near-infrared integral field data of these two galaxies, taken with SINFONI at the Very Large Telescope, and use axisymmetric orbit models to determine the masses of the SMBHs. The SMBH mass of NGC 3368, averaged over the four quadrants, is $\langle M_\bullet \rangle = 7.5 \times 10^6$ $M_\odot$ with an error of 1.1×10^6 $M_\odot$, which mostly comes from the non-axisymmetry in the data. For NGC 3489, a solution without black hole cannot be excluded when modelling the SINFONI data alone, but can be clearly ruled out when modelling a combination of SINFONI, OASIS and SAURON data, for which we obtain $M_\bullet = (6.00^{+0.56}_{-0.54}|_{\text{stat}} \pm 0.64|_{\text{sys}}) \times 10^6$ $M_\odot$. Although both galaxies seem to be consistent with the $M_\bullet$-σ relation, at face value they do not agree with the Marconi & Hunt relation between bulge magnitude and black hole mass when the total bulge magnitude (i.e., including both classical bulge and pseudobulge) is considered; the agreement is better when only the small classical bulge components are considered. However, taking into account the ageing of the stellar population could change this conclusion.

6.1 Introduction

Bulges located in the central regions of disc galaxies are commonly identified as the region where the excess light above the outer exponential disc dominates the surface brightness profile. Bulges were generally regarded as scaled-down versions of elliptical galaxies, probably formed via minor galaxy mergers. There is now evidence that there is also a second type of central structure, the so-called pseudobulges, which are more similar to mini-discs than to mini-ellipticals. They were first noted by Kormendy (1982), and Kormendy (1993) and Kormendy & Kennicutt, Jr. (2004) review properties and formation mechanisms and present a number of examples. Pseudobulges are thought to be the result of secular evolution and can be identified e.g. through the presence of disc-like structure (nuclear spirals, bars or rings), a flattening similar to that of the outer disc, rotation-dominated kinematics, exponential surface brightness profiles or young stellar populations. As the formation mechanisms of classical and pseudobulges are fundamentally different and can happen independently, galaxies could harbour both types of bulges (Athanassoula, 2005; Erwin, 2008; Erwin et al., 2003). But this fundamental difference between the formation mechanisms also leads to the question whether and how a central black hole grows inside a pseudobulge and how the mass of the black hole relates to pseudobulge properties. Supermassive black holes (SMBHs) in elliptical galaxies and classical bulges are known to follow tight correlations with luminosity (e.g. Kormendy & Richstone 1995; Marconi & Hunt 2003), mass (Häring & Rix, 2004) and velocity dispersion ($M_\bullet$-σ relation, Ferrarese & Merritt 2000; Gebhardt et al. 2000b) of the bulge.

CHAPTER 6. THE PSEUDOBULGE GALAXIES NGC 3368 AND NGC 3489

Table 6.1: Galaxy properties of NGC 3368 and NGC 3489. As both galaxies host a composite bulge, the effective radius R_e and the bulge K-band magnitude M_K are given for the pseudobulge and the classical bulge component.

Galaxy	Type	D (Mpc)	PA (°)	i (°)	R_e^{PB} (″)	M_K^{PB}	R_e^{CB} (″)	M_K^{CB}	Activity[a]
NGC 3368	SAB(rs)ab	10.4	172	53	24.9	-23.42	1.6	-19.48	L2
NGC 3489	SAB(rs)0+	12.1	71	55	4.3	-21.91	1.3	-20.60	T2/S2

[a]Ho et al. (1997)

It is not clear whether pseudobulges follow the same relations, for several reasons: 1. There are only very few direct SMBH mass measurements in pseudobulges. 2. The concept of pseudobulges is relatively new and the classification criteria therefore differ somewhat from author to author. 3. The fact that at least some galaxies could contain both bulge types (composite bulges) makes the classification and correlation studies even more complicated. The composite bulges need to be decomposed properly in order to find out with which property of which bulge component the SMBH mass correlates. Kormendy (2001) did not find any dependence of the $M_\bullet$-σ relation on the mechanism that feeds the black hole. In contrast Hu (2008) finds that the black holes in pseudobulges have systematically lower masses than black holes in classical bulges and ellipticals with the same velocity dispersion. Both studies suffer from small number statistics and unclear classification issues. For low-mass galaxies without classical bulge (i.e. likely hosts of a pseudobulge) and with virial SMBH mass estimates Greene, Ho & Barth (2008) found no deviation from the $M_\bullet$-σ relation, but a likely disagreement with the $M_\bullet$-M_{bulge} relation. Gadotti & Kauffmann (2009) found for a large number of SDSS galaxies that pseudobulges, classical bulges and ellipticals cannot follow both the $M_\bullet$-σ and the $M_\bullet$-M_{bulge} relation at the same time. As they estimated $M_\bullet$ from the $M_\bullet$-M_{bulge} relation by Häring & Rix (2004) it is not clear whether their pseudobulges follow a different $M_\bullet$-σ or $M_\bullet$-M_{bulge} relation or both.

In this chapter we present a thorough analysis and derivation of the black hole masses via extensive stellar dynamical modelling of NGC 3368, a double-barred spiral galaxy (Erwin, 2004) of type SAB(rs)ab with a well-defined pseudobulge and a very small classical bulge component, and NGC 3489, an SAB(rs)0+ galaxy with a weak large bar, a small pseudobulge and a similar-sized classical bulge. All important parameters of the two galaxies are listed in Table 6.1. Using high-resolution imaging we are able to identify and decompose the pseudobulge and classical bulge components. High-resolution adaptive-optics assisted near-IR integral-field spectroscopy enables us to model each quadrant separately and thus assess the influence of deviations from axisymmetry

6.1. INTRODUCTION

in the velocity fields on our modelling. In contrast to our two previous studies of elliptical galaxies (Chapters 4 and 5), non-axisymmetries may play a larger role.

The nucleus of NGC 3368 is weakly active and can be classified as a LINER2 based on optical emission line ratios (Ho et al., 1997). Maoz et al. (2005) and Maoz (2007) report long-term UV variations, which suggests the presence of an AGN and thus a SMBH. NGC 3489 has a weak LINER/HII transition type or Seyfert 2 nucleus (Ho et al., 1997).

We adopt a distance to NGC 3368 of 10.4 Mpc throughout this chapter based on surface brightness fluctuation measurements (Tonry et al., 2001). At this distance, $1''$ corresponds to ~ 50 pc. For NGC 3489 we adopt a distance of 12.1 Mpc (~ 59 pc/$''$), also based on the measurements of Tonry et al. (2001). If the $M_\bullet$-σ relation (Tremaine et al., 2002) applies, both galaxies are close enough to resolve the sphere of influence of the black hole from the ground with adaptive optics.

This chapter is organised as follows: In Section 6.2 we discuss the morphology of the two galaxies, including photometric (and spectroscopic) evidence for pseudobulges and classical bulges. The spectroscopic data, including stellar kinematics, gas kinematics and line strength indices are described in Section 6.3. The stellar dynamical modelling procedure and the results for the SMBH mass of each galaxy are presented in Section 6.4 and Section 6.5. Section 6.6 summarises and discusses the results.

6.2 Imaging[2]

6.2.1 Imaging data and calibrations

The imaging data we use comes from a variety of sources, including the Two-Micron All-Sky Survey (2MASS; Skrutskie et al., 2006), the Sloan Digital Sky Survey (SDSS; York et al., 2000), and the *HST* archive. We also use near-IR images taken with INGRID (a 1024^2 near-IR imager with $0.24''$ pixels) on the William Herschel Telescope (WHT): a K-band image of NGC 3368 from Knapen et al. (2003), available via NED, and an H-band image of NGC 3489 obtained during service/queue time (February 11, 2003). The seeing for the INGRID images was $0.77''$ FWHM for NGC 3368 and $0.74''$ FWHM for NGC 3489. Finally, we also use K-band images created from our SINFONI datacubes (see Fig. 6.13).

The SDSS r-band images are used for measuring the shape of the outer-disc isophotes, which helps us determine the most likely inclination for each galaxy. The *HST* archival images are used to help determine the innermost isophote shapes and surface brightness profiles. For NGC 3368, we use a NICMOS2 F160W image (PI Mulchaey, proposal ID 7330); we also use F450W and F814W WFPC2 images (PI Smartt, proposal ID 9042) to construct high-resolution colour maps and attempt to correct the NICMOS2 image for dust extinction. For NGC 3489, we used F555W and F814W WFPC2 images (PI Phillips, proposal ID 5999). We attempted to correct the NICMOS2 F160W image (for NGC 3368) and the WFPC2 F814W image (for NGC 3489) for dust extinction, following the approach of Barth et al. (2001) and Carollo et al. (1997). This involved creating a $V-H$ colourmap for NGC 3368 and a $V-I$ colourmap for NGC 3489, then generating corresponding A_H and A_I extinction maps and correcting the NICMOS2 and WFPC2 F814W images. The results were reasonably successful for NGC 3489, but less so for NGC 3368, perhaps due to the much stronger extinction in the latter galaxy.

The 2MASS images are used primarily to calibrate the INGRID near-IR images. Since the latter suffer from residual sky-subtraction problems, we calibrate them by matching surface-brightness profiles from the INGRID images with profiles from the appropriate 2MASS images (K-band for NGC 3368, H-band for NGC 3489), varying both the scaling and a constant background offset until the differences between the two profiles are minimized. We then carry over this calibration to surface-brightness profiles from the *HST* images: i.e., we calibrate the NICMOS2 F160W profile

---

[2]Based on service observations made with the William Herschel Telescope operated on the island of La Palma by the Isaac Newton Group in the Observatory del Roque de los Muchachos of the Instituto de Astrofísica de Canarias.

6.2. IMAGING

to K-band for NGC 3368 by matching it to the (calibrated) INGRID K-band profile (including a background offset), and similarly match the WFPC2 F814W profile to INGRID H-band profile for NGC 3489. Profiles from the SINFONI K-band images are then calibrated by matching them to the appropriate calibrated *HST* profiles.

In the following we will use the term "photometric bulge" to indicate the central region, where the excess light above the outer exponential disc dominates the surface brightness profile. The term "pseudobulge" then refers to the part of the photometric bulge that shows signatures of a cold, disc-like component, and the term "classical bulge" in turn refers to the part of the photometric bulge, that shows signatures of a hot elliptical-like component.

6.2.2 NGC 3368

Morphological overview and evidence for a composite pseudobulge

NGC 3368 is a relatively complex spiral galaxy, with a number of different stellar components. Erwin (2004) argued that the central regions of NGC 3368 included at least three distinct components: an outer bar with semi-major axis $a \approx 61-75''$ (4.4–5.4 kpc, deprojected), an "inner disc" extending to $a \approx 21-30''$ (1.1–1.6 kpc, deprojected), and an inner bar with $a \approx 3.4-5.0''$ (200–300 pc, deprojected). This set of nested structures is rather similar to that of the double-barred galaxy NGC 3945 (Erwin & Sparke, 1999), where Erwin et al. (2003) found that the galaxy's "photometric" bulge could be decomposed into a bright, kinematically cool disc (first noted by Kormendy 1982) with an exponential profile and a much smaller, rounder object dominating the inner few hundred parsecs – apparently a central, spheroidal bulge. Fig. 6.1 shows a global bulge-disc decomposition for NGC 3368, in which the photometric bulge (the Sérsic component of the fit) dominates the light at $r < 50''$; thus, as in NGC 3945, the "inner disc" corresponds to the photometric bulge.

In Fig. 6.2 we use the long-slit kinematic data of Héraudeau et al. (1999) to show an estimate of the local ratio of ordered to random stellar motions as a function of radius: V_{dp}, which is the observed stellar velocity deprojected to its in-plane value (assuming an axisymmetric velocity field), divided by the observed velocity dispersion at the same radius. Even though the data are all inside the photometric bulge ($r < 50''$), the ratio of V_{dp}/σ rises above 1 over much of this region. This is certainly higher than one would expect for a classical (kinematically hot) bulge, in which stellar motions are dominated by velocity dispersion. An unpublished spectrum with higher S/N from the Hobby-Eberly Telescope (M. Fabricius, private communication) shows even larger values of V_{dp}/σ for $r > 30''$, as well as $V_{dp}/\sigma > 1$ on both sides of the centre at $r \sim 5-9''$. Our tentative conclusion is that most of the photometric bulge is thus a discy, kinematically cool pseudobulge,

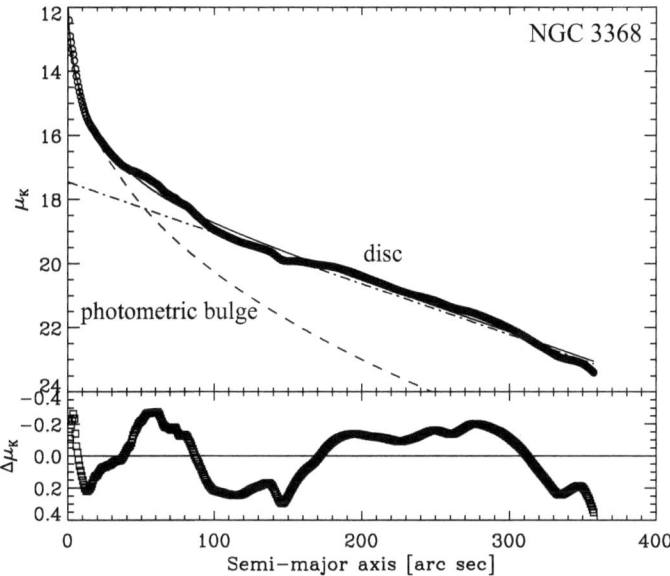

Figure 6.1: Global bulge-disc decomposition of NGC 3368. The data points (circles) are the major-axis K-band profile, combining the *HST* NICMOS2 F160W image ($r < 7.1''$) with the K-band image of Knapen et al. (2003) for $7.2 < r < 86''$ and the SDSS r-band image for $r > 86''$, all calibrated to K. Also shown is the best Sérsic + exponential fit to the data and the residuals (bottom panel). The Sérsic component represents the "photometric bulge," which dominates the light at $r < 50''$.

6.2. IMAGING

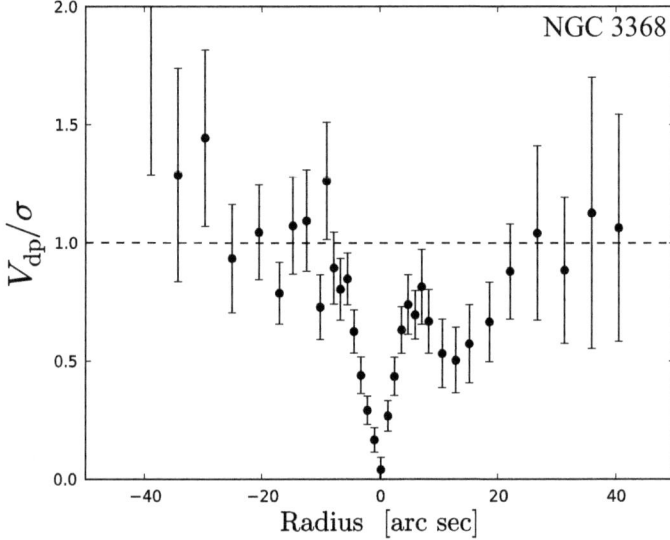

Figure 6.2: Local estimates of the ratio of ordered motion to random motion in stellar kinematics along the major axis of NGC 3368, within the photometric bulge region ($r < 50''$), based on the long-slit data of Héraudeau et al. (1999). We first deproject the stellar velocities to their in-plane values (correcting from the observed PA of 5° to our adopted major-axis PA of 172°), then divide them by the observed velocity dispersion values. Ratios < 1 indicate a dominance of velocity dispersion over bulk rotation; ratios > 1 indicate kinematically cooler regions.

similar to that found in NGC 3945. We note that the photometric bulge of NGC 3368 has also been classified as a pseudobulge by Drory & Fisher (2007), based on morphological features in the *HST* images.

The ellipse fits to ground-based and *HST* near-IR images (Fig. 6.3) suggest that NGC 3368 also harbours a small classical bulge, again in analogy to NGC 3945. Inspection of the isophotes (Fig. 6.4) shows that the inner-disc region, where the ellipticity rises to a local maximum of ~ 0.3 and the isophotes are closely aligned with the outer disc, is dominated by an elliptical structure with an exponential surface-brightness profile, extending to $r \sim 12''$, in which the inner bar is embedded (panel d of Fig 6.4). Inside the inner bar (semi-major axis $a < 2''$), the isophotes become quite round, with a mean ellipticity of ≈ 0.1. The isophotes in this region also twist significantly; inspection of both the NICMOS2 image and our SINFONI datacubes indicate that this twisting is

CHAPTER 6. THE PSEUDOBULGE GALAXIES NGC 3368 AND NGC 3489

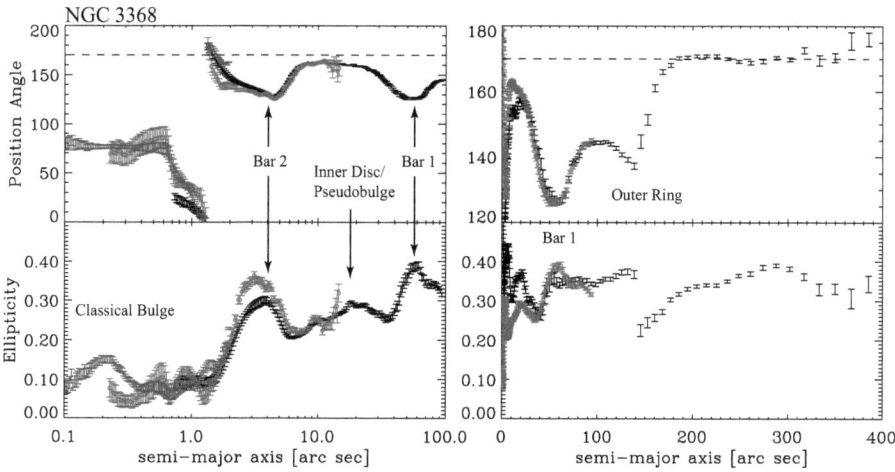

Figure 6.3: Isophotal ellipse fits for NGC 3368, plotted on logarithmic (left panel) and linear (right panel) scales. Black points are for the SDSS r-band image, while red points are for the K-band image of Knapen et al. (2003) and purple points are for the NICMOS2 F160W image. Major morphological features are marked ("Bar 1" = outer bar, "Bar 2" = inner bar). The dashed line in the top panels indicates our adopted position angle (172°) for the galaxy disc.

produced by strong dust lanes on either side of the galaxy centre. The true (unextincted) ellipticity in this region is probably close to 0.

Fig. 6.5 shows a Sérsic + exponential decomposition of the inner $r < 10''$ region. (Note that in this decomposition, we are now treating what we previously identified as the photometric bulge as simply the inner part of the disc, which has a relatively steep inner exponential profile, plus a central Sérsic excess.) The result is a surprisingly good fit, suggesting that the inner $r < 2''$ region – where, as noted, the isophotes are quite round – is a separate component (best-fit Sérsic parameters: $n = 2.35$, $R_e = 1.60''$, $\mu_e = 14.53$).

Orientation and inclination of the galaxy

Our ellipse fits of the merged SDSS r-band image (black points in Fig. 6.3) shows a consistent position angle of $\approx 172°$ for the outermost isophotes. These ellipse fits extend well outside the star-forming outer ring ($r \sim 180''$) and are thus unlikely to be affected by any intrinsic noncircularity of the ring itself. This position angle agrees very well with the *kinematic* position angles deter-

6.2. IMAGING

NGC 3368

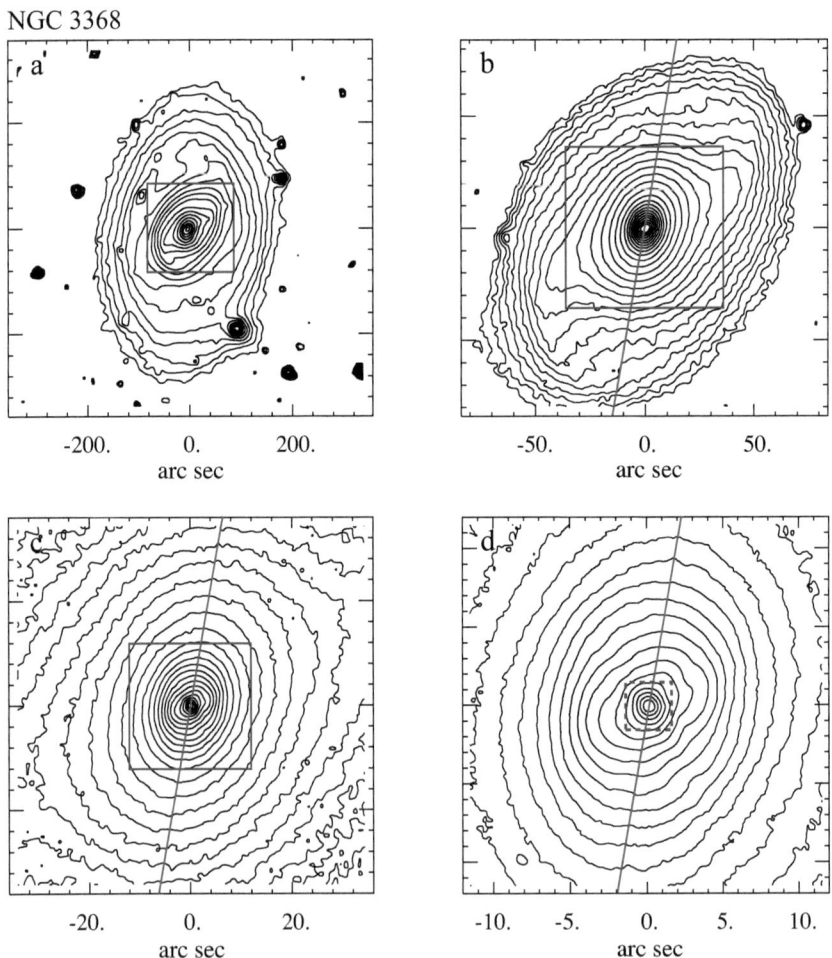

Figure 6.4: Isophotal maps of NGC 3368 on different scales, all with logarithmic intensity scaling. (a) SDSS r-band isophotes, showing the outer disc. The gray box outlines the region shown in the next panel. (b) K-band isophotes from the image of Knapen et al. (2003), showing the outer bar and the lens just outside. (c) K-band isophotes, showing the region inside the outer bar, including the bright inner pseudobulge region. (d) K-band isophotes, showing the elliptical contours of the inner pseudobulge, the inner bar, and the central bulge inside. The dashed blue square shows the approximate field of view of our 100mas SINFONI observations. For reference, we indicate the adopted major-axis position angle (172°) with the diagonal red line in panels b-d.

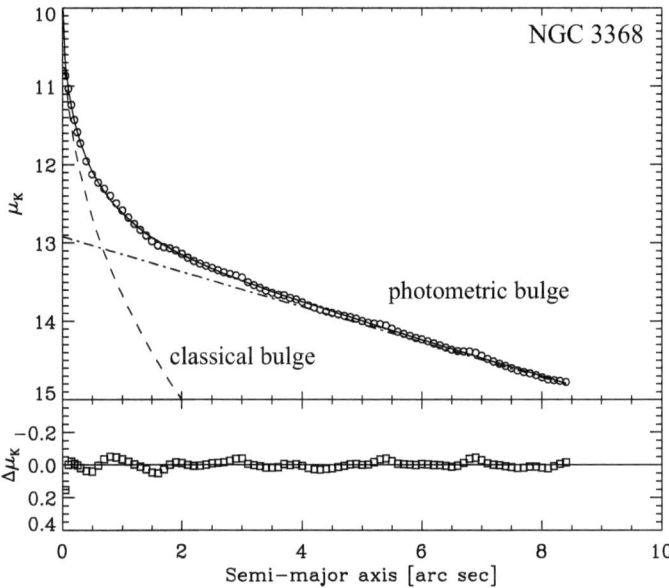

Figure 6.5: Bulge-disc decomposition of the inner photometric bulge of NGC 3368. The data points (circles) are the major-axis K-band profile for the inner $r < 8.5''$, combining our SINFONI data cube ($r < 0.3''$) with the *HST* NICMOS2 F160W image, both calibrated to K. Also shown is the best Sérsic + exponential fit to the data at $r \leqslant 8''$ and the residuals (bottom panel); in this fit, the Sérsic component represents the "classical bulge", while the exponential is the inner part of the photometric bulge.

6.2. IMAGING

mined from both H I observations (PA ≈ 170°, based on the data of Schneider 1989, as reported by Sakamoto et al. 1999) and from the Fabry-Perot Hα+[N II] velocity field of Sil'chenko et al. (2003, see also Moiseev, Valdés & Chavushyan 2004). Sil'chenko et al. also find a kinematic position angle of 170 − 175° in the stellar kinematics of the inner 2 − 5″, from their IFU data.

The ellipticity of the outer r-band isophotes is ≈ 0.37, with a range of 0.34 − 0.39. A lower limit on the inclination is thus 51°, for a razor-thin disc; thicker discs imply higher inclinations. For an intrinsic thickness of $c/a = 0.2 - 0.25$, the inclination is $i \approx 53°$. This is close to the inclination of 50° estimated by Barberà, Athanassoula & García-Gómez (2004), based on Fourier analysis of Frei et al. (1996) images (note that these images do not extend beyond the outer-ring region, and so they might in principle be biased if the ring is noncircular).

An additional, independent estimate of the inclination can be had by inverting the Tully-Fisher relation: since we know the observed H I velocity width *and* the distance to NGC 3368, we can determine the inclination needed to make the galaxy follow the Tully-Fisher relation. We use the recently published 2MASS Tully-Fisher relation of Masters et al. (2008). For a 2MASS K-band "total" magnitude (Jarrett et al., 2003) of 6.31 (including a slight reddening correction from Schlegel et al. 1998, as given by NED) and a distance of 10.4 Mpc, the absolute magnitude is $M_K = -23.795$. Using the "Sb" T-F relation from table 3 of Masters et al. (2008), this corresponds to a corrected, edge-on velocity width of $W_{\rm corr} = 425$ km s^{-1}. For the observed width, we use the tabulated value in Springob et al. (2005), which is $W_c = 324$ km s^{-1}; after applying the recommended correction for turbulent broadening (6.5 km s^{-1}), this becomes $W_{\rm corr} \sin i = 317.5$ km s^{-1}, and thus $i = 48°$.

Taken all this into considerations, we can argue that NGC 3368 has a line of nodes with PA ≈ 172° and an inclination somewhere between 48° and 55°, most likely ≈ 53°.

6.2.3 NGC 3489

Morphological overview and evidence for a composite pseudobulge

NGC 3489 is structurally somewhat simpler than NGC 3368, with only one bar instead of two. The bar itself is rather weak and difficult to recognise, because it lies almost along the minor axis of the galaxy. Projection effects thus foreshorten it so that it is visible primarily due to the abrupt isophote twists, manifesting in the ellipse fits as an extremum in the fitted position angle and a *minimum* in the ellipticity (Fig. 6.6 and Fig. 6.7). Further outside, the isophotes become maximally elongated at $r \sim 50″$, and then converge to a mean ellipticity of ≈ 0.41 at larger radii.

As shown by Erwin & Sparke (2003), the ellipticity peak at $r \sim 50''$ is due to an outer ring; the lower ellipticity outside is thus the best representation of the outer disc.

We do find some evidence for a composite pseudobulge-plus-classical-bulge structure in NGC 3489, as in NGC 3368. Specifically, while a global bulge-disc decomposition (Fig. 6.8) yields a Sérsic component which dominates the light at $r \leqslant 10''$, the isophotes interior to the bar are still fairly flattened (e.g., ellipticity ≈ 0.33 at $r \sim 6-8''$), while the central isophotes have ellipticity < 0.2.

There is in addition kinematic evidence that the photometric bulge is partly a pseudobulge. In Fig. 6.9 we plot the local ratio of (deprojected) stellar rotation velocity to velocity dispersion. These values are based on synthesized long-slit profiles derived from the SAURON velocity and velocity dispersion fields (Emsellem et al., 2004), using a $0.94''$ slit at PA = $71°$. The V_{dp}/σ ratio rises to values > 1 at $r \geqslant 7''$, still within the photometric bulge-dominated region, which suggests that the photometric bulge is at least partly dominated by rotation.

As in the case of NGC 3368, we find that the photometric bulge can be decomposed into an exponential plus a smaller Sérsic component (Fig. 6.10). The Sérsic component dominates the light at $r \leqslant 2''$, where V_{dp}/σ is clearly < 1. Note, however, that the isophotes do become quite elliptical at $r \sim 1''$, so we do not have as clean a case as in NGC 3368 for a rounder spheroidal component.

Orientation and inclination of the galaxy

Lacking the extensive large-scale gas kinematic information that was available for NGC 3368, we rely on the isophotes of the outer disc to determine the global orientation of NGC 3489. Fortunately, apart from the local maximum in ellipticity at $r \sim 50''$ due to the outer ring (see above), the outer disc is fairly well defined, with position angle = $71°$ and a mean ellipticity of 0.41, corresponding to an inclination of $55°$.

6.3 Spectroscopy[3]

6.3.1 Data and data reduction

NGC 3368 and NGC 3489 were observed between March 22 and 24, 2007, as part of guaranteed time observations with SINFONI (Bonnet et al., 2004; Eisenhauer et al., 2003a), an adaptive-optics assisted integral-field spectrograph at the VLT UT4. We used the K-band grating and the $3 \times 3''$ field

---

[3]Based on observations at the European Southern Observatory (ESO) Very Large Telescope (VLT) [078.B-0103(A)].

6.3. SPECTROSCOPY

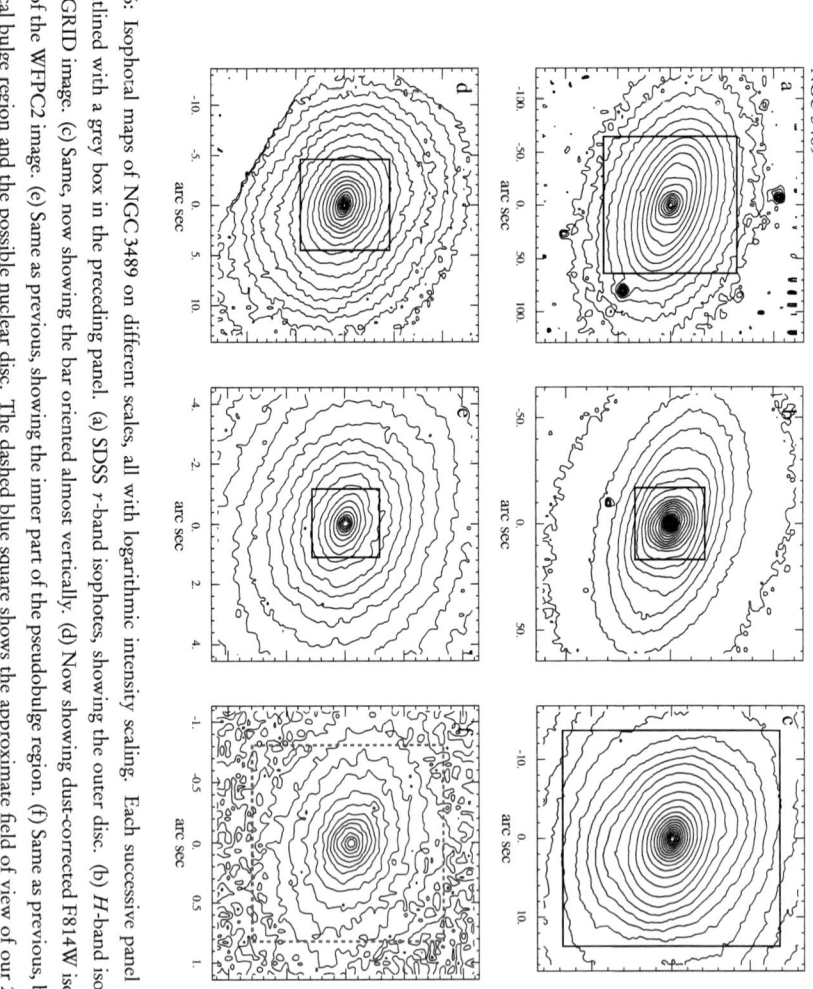

Figure 6.6: Isophotal maps of NGC 3489 on different scales, all with logarithmic intensity scaling. Each successive panel is a zoom of the region outlined with a grey box in the preceding panel. (a) SDSS r-band isophotes, showing the outer disc. (b) H-band isophotes from our WHT-INGRID image. (c) Same, now showing the bar oriented almost vertically. (d) Now showing dust-corrected F814W isophotes from the PC chip of the WFPC2 image. (e) Same as previous, showing the inner part of the pseudobulge region. (f) Same as previous, but now showing the classical bulge region and the possible nuclear disc. The dashed blue square shows the approximate field of view of our 25mas SINFONI observations.

CHAPTER 6. THE PSEUDOBULGE GALAXIES NGC 3368 AND NGC 3489

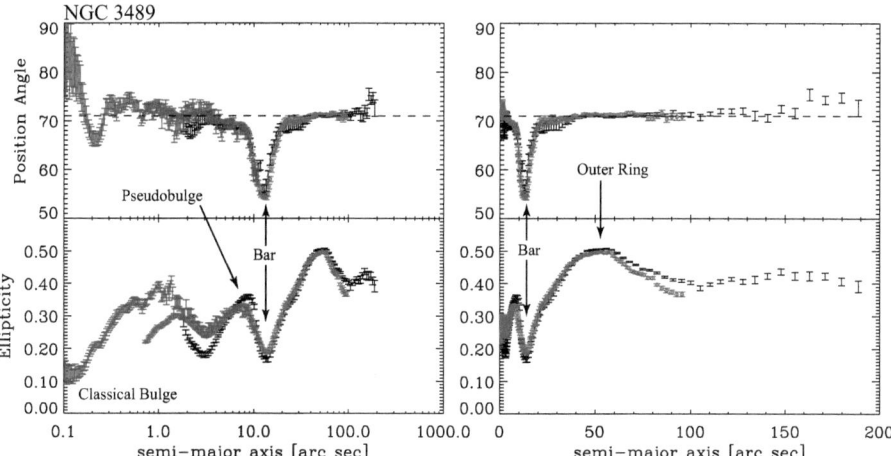

Figure 6.7: Isophotal ellipse fits for NGC 3489, plotted on logarithmic (left panel) and linear (right panel) scales. Black points are for the SDSS r-band image, while red points are for the INGRID H-band image and purple points are for the dust-corrected WFPC2 F814W image. Major morphological features are indicated. Note that due to projection effects, the bar shows up as both a strong twist in the position angle and a *minimum* in the ellipticity. The dashed line in the top panels indicates our adopted position angle (71°) for the galaxy disc.

6.3. SPECTROSCOPY

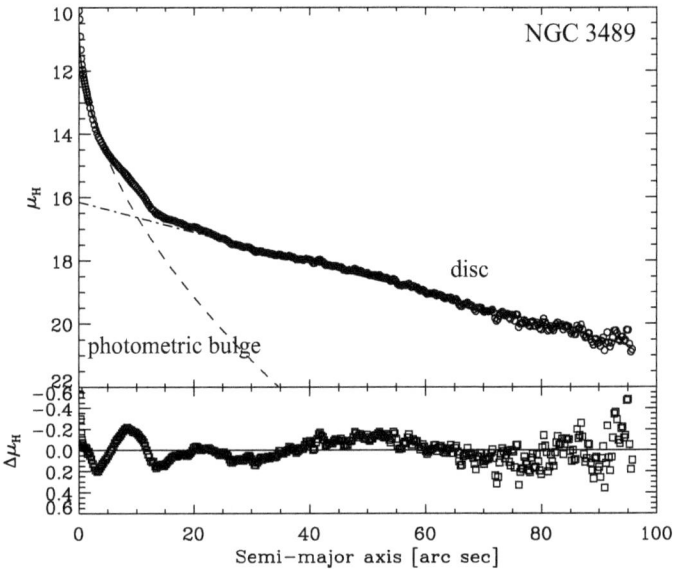

Figure 6.8: Global bulge-disc decomposition of NGC 3489. The data points (circles) are the major-axis cut, combining our ground-based H-band image ($r > 1.7''$) with the dust-corrected WFPC2 F814W image (scaled to H-band). Also shown is the best Sérsic + exponential fit to the data and the residuals (bottom panel). The Sérsic component represents the "photometric bulge," which dominates the light at $r < 10''$.

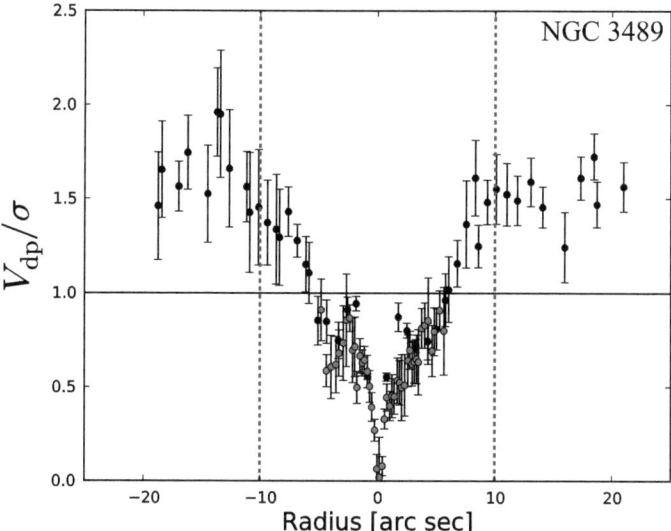

Figure 6.9: Local estimates of the ratio of ordered motion to random motion in stellar kinematics along the major axis of NGC 3489, using major-axis profiles extracted from the velocity and velocity dispersion fields of the SAURON (black) and OASIS (green) observations. Stellar velocities are deprojected to their in-plane values, then divided by the observed velocity dispersion values. Ratios < 1 indicate a dominance of velocity dispersion over bulk rotation. At $r \leqslant 10''$ (vertical dotted line), the photometric bulge dominates the light (see Fig. 6.8). Since $V_{dp}/\sigma > 1$ within this region, the outer part of the photometric bulge is still dominated by rotation, making it a kinematic pseudobulge.

6.3. SPECTROSCOPY

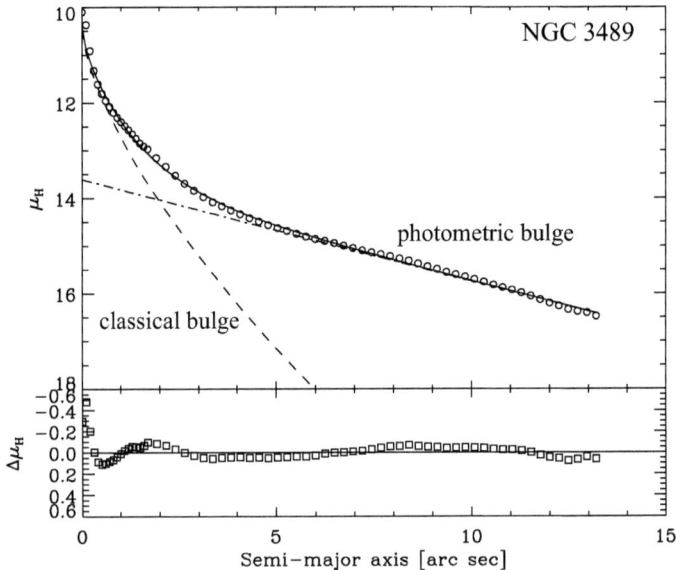

Figure 6.10: Bulge-disc decomposition of the inner photometric bulge of NGC 3489. The data points (circles) are the H-band profile from Fig. 6.8 for the inner $r < 13''$ (with data at $r < 1.7''$ coming from the dust-corrected WFPC2 F814W image). Also shown is the best Sérsic + exponential fit to the data at $r \leqslant 13''$ and the residuals (bottom panel), with the Sérsic component representing the classical bulge.

of view ($0.05 \times 0.1''$ spaxel^{-1}) for NGC 3368 and the $0.8 \times 0.8''$ field of view ($0.025 \times 0.0125''$ spaxel^{-1}) for NGC 3489. The total on-source exposure time was 80 min for NGC 3368 and 120 min for NGC 3489, consisting of 10 min exposures taken in series of "object-sky-object" cycles, dithered by a few spaxels.

The laser guide star (LGS) PARSEC (Bonaccini et al., 2002; Rabien et al., 2004) was used for the AO correction of NGC 3368, with the tip-tilt sensor closed on the nucleus of NGC 3368 ($R = 13.58$, $B - R = 1.86$ within a $3''$ diameter aperture). Although the nucleus itself is just bright enough to be used as natural guide star, its shape is rather irregular and not pointlike in the R-band due to the large amounts of dust in the nuclear regions and the lack of a strong AGN. Therefore a better AO correction was expected from using the LGS instead. The ambient conditions were good and stable, with an average seeing of $\approx 0.6''$ in the near-IR. The point-spread function (PSF) was derived by taking an exposure of a nearby star with approximately the same R-band magnitude and $B - R$ colour as the nucleus of NGC 3368, using the LGS with the PSF star itself as tip-tilt reference star. The FWHM of the PSF is $\approx 0.165''$ (see left panel of Fig. 6.11) and the achieved Strehl ratio is ≈ 0.14. Due to the time gap between the observations of the galaxy and the PSF star, the measured PSF shape could be different from the PSF during the galaxy observations. We compared the surface brightness profile of the SINFONI data with the surface brightness profile of an *HST* NICMOS2 F160W image, convolved with Gaussians of different widths and found that the NICMOS surface brightness profile most closely resembles the SINFONI profile for a FWHM ≈ 0.165 (see right panel of Fig. 6.11), confirming our PSF measurement.

NGC 3489 was observed using its nucleus with $R = 13.22$ ($3''$ diameter aperture) as natural guide star (NGS) for the AO correction. A PSF star with a similar magnitude and $B - R$ colour as the NGS was observed regularly in order to determine the spatial resolution. The ambient conditions were excellent and stable with a seeing around $0.5''$ in the near-IR, resulting in a FWHM of the PSF of $\approx 0.08''$ and a Strehl ratio of 0.43 (see Fig. 6.12).

The data reduction was done using the SINFONI data reduction package SPRED (Abuter et al., 2006; Schreiber et al., 2004) as explained in Chapter 5 (Nowak et al., 2008). The reduction of the telluric standard and the PSF reference star was done with the ESO pipeline. For the flux calibration we used the telluric standard stars Hip 046438 and Hip 085393 with 2MASS K_s magnitudes of 7.373 and 6.175 respectively as a reference. Fig. 6.13 shows the flux-calibrated images of the two galaxies, collapsed along the wavelength direction.

6.3. SPECTROSCOPY

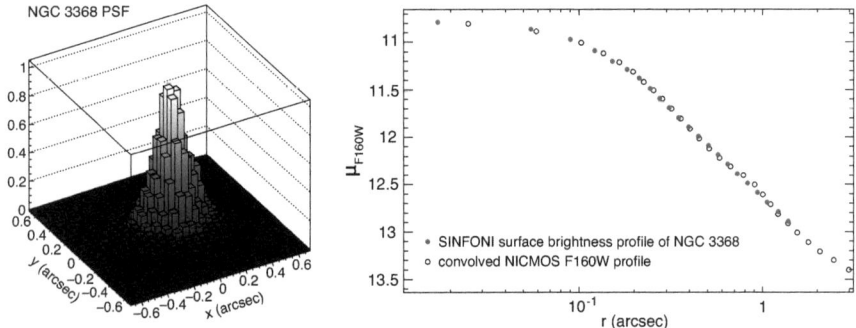

Figure 6.11: Left panel: SINFONI PSF derived by observing a star of the same magnitude and colour as the nucleus of NGC 3368. Its FWHM is 0.165″. Right: Comparison of the SINFONI K-band surface brightness profile with the surface brightness profile of an *HST* NICMOS2 F160W image convolved with a Gaussian such that the spatial resolution is 0.165″. The SINFONI profile is shifted such that it matches the NICMOS profile.

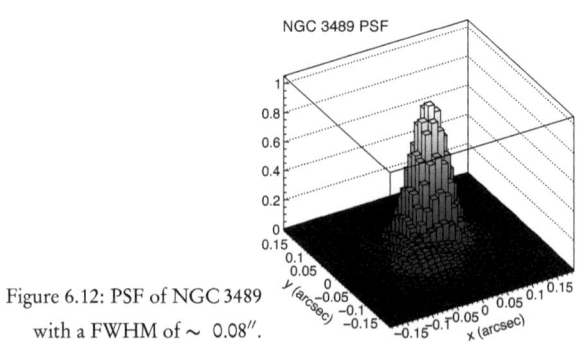

Figure 6.12: PSF of NGC 3489 with a FWHM of $\sim 0.08″$.

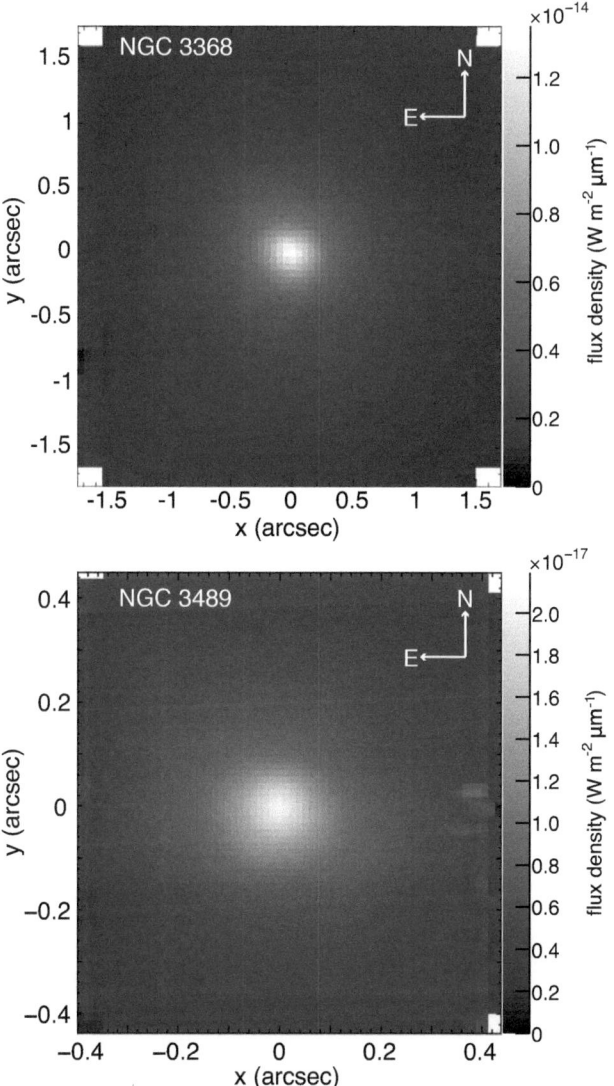

Figure 6.13: SINFONI images of NGC 3368 and NGC 3489.

6.3.2 Stellar kinematics in NGC 3368

The SINFONI data of NGC 3368 were binned using a binning scheme with five angular and ten radial bins per quadrant, adopting a major-axis position angle of 172°. As in Chapters 4 and 5 we used the maximum penalised likelihood (MPL) technique of Gebhardt et al. (2000a) to extract the stellar kinematics from the first two CO bandheads ^{12}CO(2 − 0) and ^{12}CO(3 − 1), i.e. the spectral range between 2.279 µm and 2.340 µm rest frame wavelength. With the MPL method, non-parametric line-of-sight velocity distributions (LOSVDs) are obtained by convolving an initial binned LOSVD with a linear combination of template spectra. The residual differences between the resulting model spectrum and the observed galaxy spectrum are then calculated. Then the velocity profile and the template weights are successively adjusted in order to optimise the fit by minimizing the function $\chi_P^2 = \chi^2 + \alpha \mathcal{P}$, where α is the smoothing parameter and $\mathcal{P}$ is the penalty function. The S/N of the binned spectra ranges between 80 and 120 with a mean value of $\sim$ 110. We determined the optimal smoothing parameter from the simulations in Section 3.4.2. For a galaxy with a velocity dispersion around 100 km s^{-1}, a velocity bin width of $\sim$ 35 km s^{-1} and the mentioned S/N a smoothing parameter $\alpha \approx 5$ is appropriate. As kinematic template stars we chose four K and M giants which have about the same intrinsic CO equivalent width (EW) as the galaxy (12 − 14 Å, using the EW definition and velocity dispersion correction from Silge & Gebhardt 2003). The uncertainties on the LOSVDs are estimated using Monte Carlo simulations (Gebhardt et al., 2000a). First, a reference galaxy spectrum is created by convolving the template spectrum with the measured LOSVD. Then 100 realizations of that initial galaxy spectrum are created by adding appropriate Gaussian noise. The LOSVDs of each realization are determined and used to specify the confidence intervals.

For illustration purposes we fitted Gauss-Hermite polynomials to the LOSVDs. Fig. 6.14 shows the two-dimensional fields of v, σ and the higher-order Gauss-Hermite coefficients h_3 and h_4, which quantify the asymmetric and symmetric deviations from a Gaussian velocity profile (Gerhard, 1993; van der Marel & Franx, 1993). The major-axis profiles are shown in Fig. 6.21.

The velocity field of NGC 3368 shows a regular rotation about the minor axis. The average, luminosity-weighted σ within the total SINFONI field of view is 98.5 km s^{-1}. A central σ-drop of 7% is present within the inner $\sim$ 1″, well inside the region of the classical bulge component. σ-drops are not uncommon in late-type galaxies and are usually associated with nuclear discs or star-forming rings (e.g. Comerón et al. 2008; Peletier et al. 2007; Wozniak et al. 2003). These could be formed e.g. as a result of gas infall and subsequent star formation, but as no change in ellipticity is found in the centre, such a disc would have to be very close to face-on. A σ-drop does not imply the absence of a SMBH if the centre is dominated by the light of a young and kinematically cold

CHAPTER 6. THE PSEUDOBULGE GALAXIES NGC 3368 AND NGC 3489

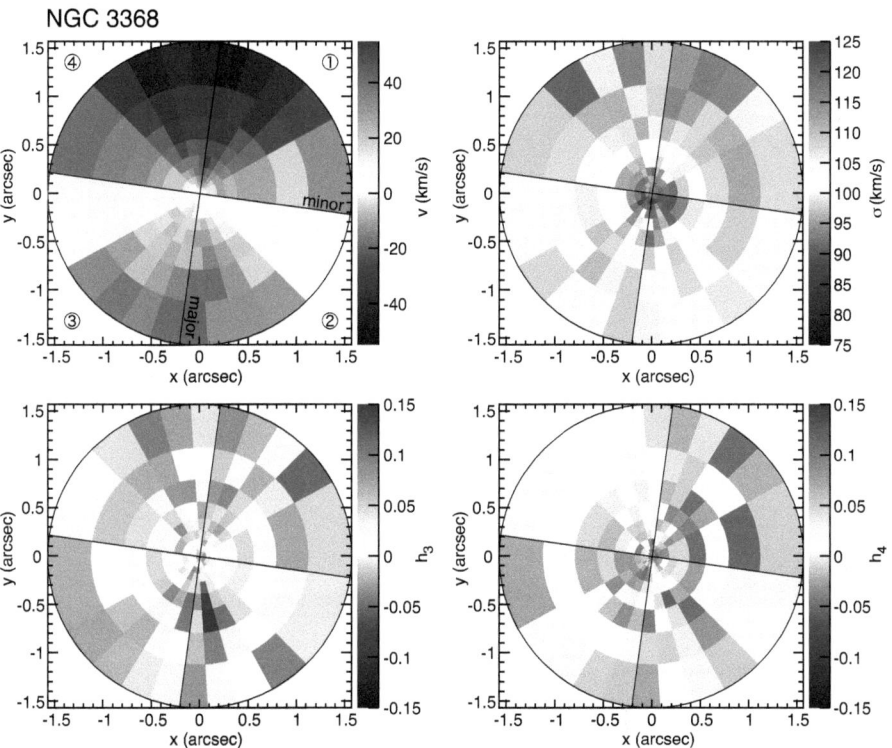

Figure 6.14: Two-dimensional stellar kinematics (v, σ, h_3 and h_4) of NGC 3368. Major axis, minor axis and the numbering of the quadrants are indicated in the velocity map (upper left).

6.3. SPECTROSCOPY

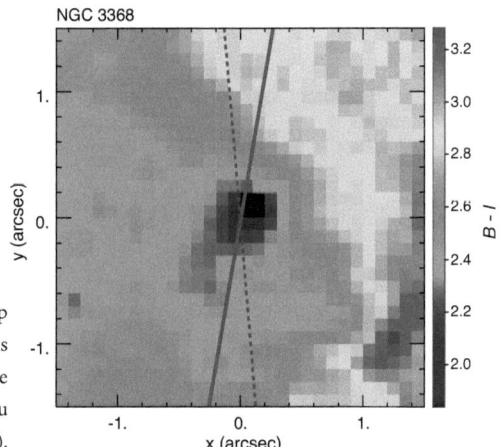

Figure 6.15: *HST* WFPC2 $B - I$ colour map of NGC 3368. Indicated is the major axis as a solid line (PA= 172°) along with the PA= 5° slit orientations used by Héraudeau et al. (1999) and Vega Beltrán et al. (2001).

stellar population. Davies et al. (2007) observed σ-drops in a number of strongly active galaxies. In these AGN the mass of the central stellar component was $\sim$ 10 times that of the SMBH, so no outstanding kinematic signature would be expected. Another example is the velocity dispersion of the Milky Way, which apparently drops in the central 100 pc, and only rises in the inner 1−2 pc (see figure 9 of Tremaine et al. 2002). Finally, a central σ-drop has been found in NGC 1399 (Gebhardt et al., 2007; Lyubenova et al., 2008), where it has been interpreted as a signature of tangential anisotropy.

The velocity dispersion in quadrants 2 and 3 is smaller than in quadrants 1 and 4. A possible explanation for that behaviour could be the substantial amounts of dust in the central regions (Fig. 6.15), although the effect of the dust in the K-band is relatively weak. The *HST* WFPC2 $B - I$ colour map (Fig. 6.15) shows that within the SINFONI field of view the dust extinction is largest in quadrants 1 and 2. Quadrant 4 is moderately affected while quadrant 3 seems to be relatively dust-free. We will further discuss the asymmetries in Section 6.4.4.

In the near-IR the presence of dust should have a much smaller effect on the kinematics than in the optical, therefore the asymmetry should be much stronger in the kinematics measured using optical absorption lines, if dust is responsible for the asymmetry. Longslit kinematics (v and σ) at PA= 5°, measured from optical spectra using Fourier-Fitting or FCQ (Bender, Saglia & Gerhard, 1994), are available from Héraudeau et al. (1999) and Vega Beltrán et al. (2001). Two-dimensional kinematics have been measured by Sil'chenko et al. (2003) (see also Moiseev et al. 2004) with the Multi-Pupil Field Spectrograph (MPFS) at the Russian 6 m telescope in the optical using a cross-correlation technique. The spatial resolution of the optical data is between 1.4″ and 3.0″. The velocities of the different authors are in good agreement with each other and with the SINFONI velocities considering the different seeing values. The optical velocity dispersions are, however, significantly larger than those measured with SINFONI. They are on average around 130 km s^{-1} for the longslit data and $\sim$ 150 km s^{-1} for the MPFS data. There are a number of possible causes for such a discrepancy. The authors used different correlation techniques, slightly different wavelength regions and different templates. A difference between optical and K-band σ measurements was also found by Silge & Gebhardt (2003) for a sample of galaxies and they suggested that this might be caused by strong dust extinction in the optical. But weak emission lines could also alter the absorption lines and thus the measured kinematics. As in the SINFONI data, a velocity dispersion asymmetry is also present in all optical data sets, as well as a velocity asymmetry. The velocity dispersion of Moiseev et al. (2004) is enhanced in the entire region west of the major axis, where also the majority of the dust is located (Fig. 6.15). However, when comparing the extinction along the location of the longslits of Héraudeau et al. (1999) and Vega Beltrán et al. (2001) with the according velocity dispersion, there seems to be no correlation. Thus it is not clear whether and in what way dust influences the velocity dispersion in NGC 3368.

Another explanation for the asymmetry could be lopsidedness, which is common in late-type galaxies. Possible mechanisms which could cause lopsidedness include minor mergers, tidal interactions and asymmetric accretion of intergalactic gas (Bournaud et al., 2005). As the large-scale stellar and gas velocity fields and gas distributions (Haan et al., 2008; Sil'chenko et al., 2003) are rather regular, a recent merger or collision with another galaxy seems unlikely. Accretion of gas from the intergalactic HI cloud is a more likely scenario (Schneider, 1989; Sil'chenko et al., 2003) and could be a possible explanation for the presence of molecular hydrogen clouds close to the centre (see below). However, there seems to be no lopsidedness in the K-band photometry, as any distortions of the isophotes can plausibly be explained by dust. The molecular gas distribution on the other hand is very disordered in the central $\sim$ 200 pc (see below and Haan et al. 2009). Thus if the gas mass differences between different regions of the galaxy would be large enough, they could be a plausible explanation for the distorted stellar kinematics. However, as shown later, the molecular gas mass is too small to have a significant effect on the stellar kinematics.

Central lopsidedness like an M31-like nucleus or otherwise off-centred nuclear disc (Bender et al., 2005; Jog & Combes, 2009) could, if the resolution is just not high enough to resolve the disc as such, leave certain kinematical signatures like a slightly off-centred σ-peak or σ-drop. On the other hand we see velocity asymmetries out to $r \sim 20''$, which is too large to be explained by an M31-like nuclear disc.

6.3. SPECTROSCOPY

In principle, the outer and inner bars could cause asymmetries in the stellar kinematics. However, the SINFONI field of view is located well inside the inner bar, and the only changes in velocity dispersion associated with inner bars which have been observed are symmetric and take place at the outer ends of inner bars (de Lorenzo-Cáceres et al., 2008).

6.3.3 Gas kinematics in NGC 3368

In NGC 3368 the only emission lines detected arise from molecular hydrogen H_2. The strongest line is 1-0S(1) at $\lambda = 2.1218$ μm. To determine the flux distribution and velocity of the H_2 gas we fitted a Gaussian convolved with a spectrally unresolved template profile (arc line) to the continuum-subtracted spectrum (Davies et al., 2007). The parameters of the Gaussian are adjusted such that they best fit the data. Fig. 6.16 shows the flux distribution and the velocity field of H_2 1-0S(1). As the S/N is very low in some regions we binned the data using adaptive Voronoi binning (Cappellari & Copin, 2003) to ensure a robust velocity measurement in each bin. The most striking feature seen in Fig. 6.16 are the two clouds of H_2 gas, located $\sim 0.36''$ and $\sim 0.72''$ north of the photometric centre. These two clouds are kinematically decoupled from the remaining H_2 gas distribution and seem to move in opposite directions. Their projected sizes are approximately 25 pc and 20 pc FWHM. The H_2 distribution outside these two clouds is relatively smooth. Its kinematic position angle, measured using the method described in Appendix C of Krajnović et al. (2006), is $\sim 171°$ and thus agrees with the stellar kinematic position angle. The gas velocity follows the rotation of the stars within a radius of $\sim 0.5''$; outside that radius it rotates faster, reaching rotation velocities up to ~ 100 km s^{-1}.

The origin of the H_2 clouds is unclear. H_2 emission line ratios can help to distinguish between different excitation mechanisms like shock heating, X-ray illumination or UV fluorescence (Rodríguez-Ardila, Riffel & Pastoriza, 2005). Table 6.2 shows the H_2 line ratios for different regions (the total SINFONI field of view and the two H_2 clouds). They indicate that the H_2 gas is thermalised in all regions of the SINFONI field of view (cf. figure 5 in Rodríguez-Ardila et al. 2005).

The ratio 2-1S(1)/1-0S(1) and therefore the vibrational excitation temperature (Reunanen, Kotilainen & Prieto, 2002) is smaller in the H_2 clouds than in the regions where H_2 is evenly distributed.

6.3.4 Line strength indices for NGC 3368

The stellar populations of NGC 3368 have been analysed by Sil'chenko et al. (2003) and Sarzi et al. (2005) using optical spectra. Both found that a relatively young stellar population with a mean age

CHAPTER 6. THE PSEUDOBULGE GALAXIES NGC 3368 AND NGC 3489

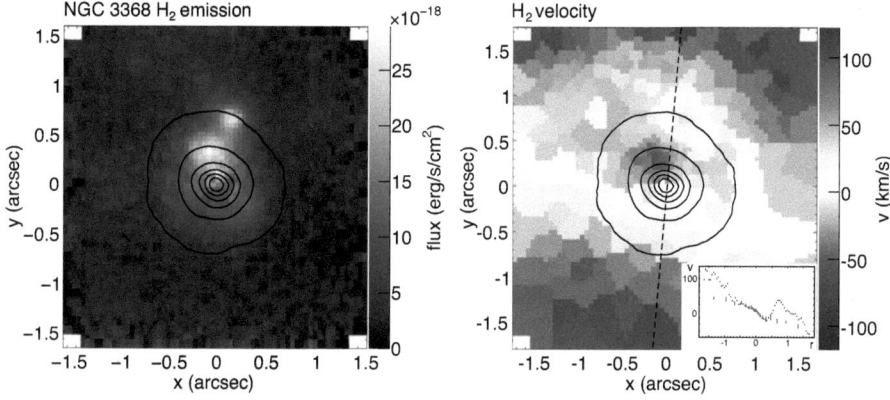

Figure 6.16: Left panel: H_2 (1 − 0S(2)) emission in the centre of NGC 3368. Right panel: H_2 gas velocity. The dashed line indicates the major axis. The small inset shows the pseudo-longslit gas velocity profile along the major axis (black points) in comparison with the major-axis stellar velocity (red points). The isophotes of the stellar emission are overlaid in both panels.

Table 6.2: H_2 1 − 0S(1) 2.12μm emission line fluxes and H_2 line ratios for the approaching gas cloud (cloud1), the receding gas cloud (cloud2) and the complete SINFONI field of view.

Region	1 − 0S(1) (10^{-15} erg/s/cm²)	$\frac{1-0S(3)}{1-0S(1)}$	$\frac{1-0S(2)}{1-0S(1)}$	$\frac{1-0S(0)}{1-0S(1)}$	$\frac{2-1S(1)}{1-0S(1)}$	$T_{\rm vib}$ (K)	$M_{H_2}^{\rm hot}$ ($M_\odot$)	$M_{H_2}^{\rm cold,a}$ ($10^7 M_\odot$)
cloud1	1.66	1.35	0.56	0.20	0.12	2280	9.13	2.23 / 0.21
cloud2	0.83	1.30	0.51	0.21	0.11	2203	4.57	1.12 / 0.10
total	21.40	1.69	0.98	0.26	0.17	2740	117.77	28.72 / 2.65

[a] The first value gives the cold H_2 mass as estimated from hot H_2 using the conversion of Mueller Sánchez et al. (2006). The second value gives the best guess for the real cold gas mass after calibration against direct mass measurements from CO (J = 1 − 0) at $r = 1.5''$ (Sakamoto et al., 1999).

6.3. SPECTROSCOPY

Table 6.3: Mean near-IR line strength indices in Å of NGC 3368 (3 × 3″ aperture) and NGC 3489 (0.8 × 0.8″ aperture). The corresponding RMS is given in brackets.

	NGC 3368	NGC 3489
Na I	4.30 (0.16)	3.30 (0.36)
Ca I	2.40 (0.29)	1.53 (0.50)
Fe I1	1.55 (0.07)	1.25 (0.23)
Fe I2	0.91 (0.10)	0.76 (0.21)
CO	17.90 (0.27)	17.70 (0.72)

of around 3 Gyr dominates the central region. Towards larger radii Sil'chenko et al. (2003) found a strong increase in age.

Fig. 6.17 shows the near-IR line indices Na I and CO measured in the same way as in Chapter 5, using the definitions of Silva et al. (2008). The average values inside the SINFONI field of view are listed in Table 6.3. They differ significantly from the relations between Na I or CO and σ found by Silva et al. (2008) for early-type galaxies in the Fornax cluster which may be, as in the case of Fornax A, probably due to the relatively young age of the stellar population. Younger populations seem to have larger Na I at equal σ than old stellar populations in the galaxy samples of Silva et al. (2008) and Cesetti et al. (2009). However, no such trend is obvious for the CO index, so the difference seen here could be due to other aspects like metallicity or galaxy formation history. The measured average indices have values which are quite similar to those found in the centre of Fornax A (Chapter 5), which could indicate that the stellar populations are quite similar in terms of age and metallicity. However, the interpretation of Na I, Ca I and Fe I must be done bearing in mind that these features always include significant contributions from other elements (Silva et al., 2008).

The radial distribution of the line indices (Fig. 6.17) shows a slight asymmetry, similar to the kinematics. The two quadrants with the smaller σ have larger CO EWs and smaller Na I EWs than the other two quadrants. In addition there seems to be a strong negative gradient in Na I and a moderately strong negative gradient in CO. Ca I and Fe I are approximately constant with radius. A small central drop is present in most indices, which could indicate the presence of weak nuclear activity (see Davies et al. 2007).

CHAPTER 6. THE PSEUDOBULGE GALAXIES NGC 3368 AND NGC 3489

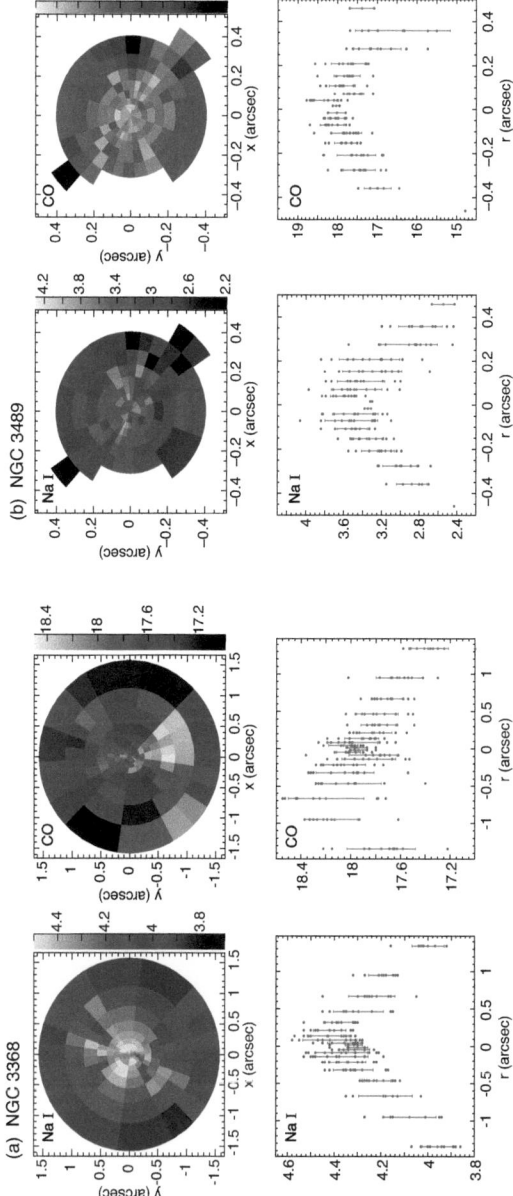

Figure 6.17: Near-IR line indices Na I and CO in Å for (a) NGC 3368 and (b) NGC 3489. The upper row shows the two-dimensional index fields, with the position-index diagrams in the bottom row. Individual values are plotted in grey, while overplotted in red are the mean indices and their RMS for bins belonging to semicircles in the receding half of the galaxy (negative radius r) or the approaching half (positive radius r).

6.3. SPECTROSCOPY

6.3.5 Stellar kinematics in NGC 3489

The NGC 3489 data were binned in the same way as the NGC 3368 data, with identical angular bins and nine somewhat smaller radial bins due to the higher spatial resolution. A position angle of 71° was used. The mean S/N is 70, for which a smoothing parameter of $\alpha \approx 8$ is appropriate. As kinematic template stars we chose four K and M giants with an intrinsic CO equivalent width in the range 13 – 15 Å.

Fig. 6.18 shows the v, σ, h_3 and h_4 maps of NGC 3489. The kinematics is similar to that of NGC 3368 in some aspects. It is clearly rotating about the minor axis, though stronger than NGC 3368. The velocity dispersion also drops towards the centre by around 4%, but then has a tiny peak in the central bins. The average σ in the total $0.8 \times 0.8''$ field of view is 91 km s^{-1}. The h_3 values clearly anticorrelate with v. h_4 is on average small and negative. Along the major axis it is positive in the outer bins and negative in the inner bins. No asymmetry is present in σ, but the velocity is asymmetric. It increases strongly on the receding side and then remains approximately constant at $r > 0.05''$, whereas on the approaching side the slope is less steep and an approximately constant velocity is reached much further out at $r > 0.2''$. The major-axis profiles are shown in Fig. 6.26. Note that the central velocity bin is omitted in that plot. Despite the presence of strong dust features in optical images, the kinematics is in comparatively good agreement with the 2D SAURON (Emsellem et al., 2004) and OASIS (McDermid et al., 2006) kinematics, though due to the high spatial resolution and the very small field of view of our data a direct comparison with seeing-limited data is not easy. The SINFONI velocity field seems to be fully consistent with the optical velocity. The average SINFONI velocity dispersion is smaller than the central SAURON and OASIS σ. The central SAURON and OASIS h_4 is significantly larger than the SINFONI values, and the anticorrelation of h_3 and v seems to be less strong in general and essentially nonexistent in the central arcsecond of the OASIS data.

Longslit kinematics is available from Caon, Macchetto & Pastoriza (2000), who used a cross-correlation technique to determine v and σ. Their σ is much larger (117 km s^{-1} in the central pixels of the major-axis longslit), which could be due to their cross-correlation technique or template mismatch. In addition the σ-profile seems to be slightly asymmetric, but only at large radii, where the errors are large.

6.3.6 Line strength indices for NGC 3489

The stellar populations of NGC 3489 in the centre have been analysed by Sarzi et al. (2005) using optical *HST* STIS longslit spectra and by McDermid et al. (2006) using OASIS integral-field data.

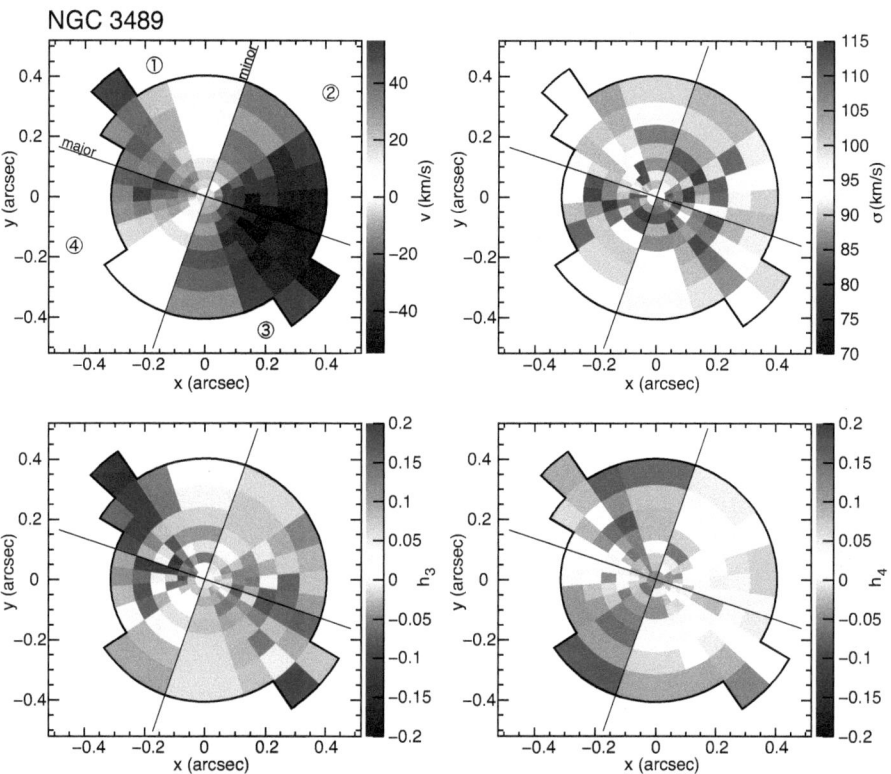

Figure 6.18: Two-dimensional stellar kinematics (v, σ, h_3 and h_4) of NGC 3489. Major axis, minor axis and the numbering of the quadrants are indicated in the velocity map (upper left).

6.4. DYNAMICAL MODELLING OF NGC 3368

Sarzi et al. (2005) obtained a mean age of about 3.1 Gyr in the central $0.2 \times 0.25''$ by fitting stellar population synthesis models to the spectra, assuming solar metallicity. McDermid et al. (2006) obtained a mean age of 1.7 Gyr in the central $8 \times 10''$ with an age gradient down to ~ 1 Gyr towards the centre from the analysis of Lick indices. These two values are more or less in agreement when taking into account the measurement errors and that the Sarzi et al. (2005) value would decrease when considering the metallicity increase to supersolar values in the centre measured by McDermid et al. (2006). Another possibility is that the central $\sim 0.2''$, which are unresolved by McDermid et al. (2006), contain an older stellar population.

The near-IR absorption-line indices Na I and CO are shown in Fig. 6.17b. They seem to be, like the stellar kinematics, axisymmetric. As in the case of NGC 3368 there is a clear negative gradient in both indices, which could mean an age or a metallicity gradient, or a combination of both. The other indices, Ca I and Fe I, are largely constant. The average line strength indices within the $0.8 \times 0.8''$ field of view are given in Table 6.3. The average CO line strength is very similar to the value found in NGC 3368. All other measured indices are slightly smaller than in NGC 3368. This seems to be generally in agreement with the results of Sarzi et al. (2005), who found similar mean ages and populations in both galaxies. A small central drop is present only in Na I, implying that nuclear activity must be extremely weak or absent.

6.4 Dynamical modelling of NGC 3368

For the dynamical modelling we make use of the Schwarzschild (1979) orbit superposition technique: first the gravitational potential of the galaxy is calculated from the stellar mass density ρ_* and trial values for the black hole mass $M_\bullet$ and the mass-to-light ratio Υ. Then an orbit library is generated for this potential and a weighted orbit superposition is constructed such that it matches the observational constraints. Finally everything is repeated for other potentials until the appropriate parameter space in $M_\bullet$ and Υ is systematically sampled. The best-fitting parameters then follow from a χ^2-analysis. The deprojected luminosity density is a boundary condition and thus is exactly reproduced, while the LOSVDs are fitted. In order to accomplish these steps, we first need to determine the stellar mass profile.

6.4.1 Construction of the stellar luminosity profile

For dynamical modelling purposes, we need an appropriate surface-brightness profile and an appropriate ellipticity profile, along with an assumption of axisymmetry. While simply using the results

CHAPTER 6. THE PSEUDOBULGE GALAXIES NGC 3368 AND NGC 3489

of ellipse-fitting may be valid for an elliptical galaxy, where the approximation that the galaxy is a set of nested, axisymmetric ellipsoids with variable axis ratio but the same position angle is often valid, a system like NGC 3368, with two bars, dust lanes, and spiral arms, is clearly more complicated. Such a complex structure also makes it important to allow a mass-to-light ratio (Υ) gradient in order to account for stellar population changes. This can be conveniently approximated by using more than one components, where each component has its own Υ.

We model the luminosity distribution of NGC 3368 as the combination of *two* axisymmetric components: a disc with fixed (observed) ellipticity = 0.37, which by design includes both inner and outer bars *and* the discy pseudobulge; and a central "classical" bulge of variable (but low) ellipticity. Thus, we assume that the bars can, to first order, be azimuthally "averaged away."

The surface brightness profile of the disc component is *not* assumed to be a simple exponential. Instead, it is the *observed* surface brightness profile of the entire galaxy outside the central bulge, out to $r = 130''$, along with an inward extrapolation to $r = 0$. We base this profile on ellipse fits with fixed ellipticity and position angle ($\epsilon = 0.37$, PA $= 172°$) to the K-band image of Knapen et al. (2003), with the inner $r < 3.7''$ based on the exponential component of our inner bulge-disc decomposition (Fig. 6.5). (Comparison of profiles from the dust-corrected NICMOS2 image and the Knapen et al. image shows that seeing affects the latter only for $r < 2''$, which is already within the region where the classical bulge affects the profile.) Inspection of both this profile and a similar fixed-ellipse profile from the NICMOS2 image shows that the classical bulge begins to affect the profile only for $r < 3.7''$. Consequently, the disc profile for $r < 3.7''$ is the inward extrapolation of the exponential component from our inner Sérsic+exponential decomposition ($\mu_0 = 12.75$, $h = 5.28''$; Figure 6.5).

To generate the classical bulge, we assume, following the inner decomposition discussed above, that the light in the inner $r < 8''$ is the combination of an inner exponential and the classical bulge (Fig. 6.5). We generate a corresponding 2D exponential model for the inner disc and subtract it from the NICMOS2 image, and then fit ellipses to the residual image, which we assume is dominated by the classical bulge. This allows for possible variations in the classical bulge's ellipticity and, perhaps more importantly, uses the observed surface brightness profile at the smallest radii, rather than an analytic fit. Finally, we generate an extension of the bulge profile out to the same outermost radius as the disc profile (i.e., well outside the NICMOS2 image) by fitting a Sérsic function to the classical-bulge profile, and assuming a constant ellipticity of 0 and the same PA as for the outer disc at large radii.

The code of Magorrian (1999) was used for the deprojection assuming that all components are axisymmetric. Both components, the disc and the classical bulge, were deprojected for an incli-

6.4. DYNAMICAL MODELLING OF NGC 3368

nation $i = 53°$ as obtained from the photometry (see Section 6.2.2), and for a few nearby values between 52° and 55°. No shape penalty was applied. The stellar mass density then can be modelled, as in Chapter 4, via $\rho_* = \Upsilon_{bulge}\nu_{bulge} + \Upsilon_{disc}\nu_{disc}$, where ν is the luminosity density obtained from the deprojection and Υ is the mass-to-light ratio in units of $M_\odot/L_\odot$. Υ is assumed to be constant with radius for each component, which is approximately true in the central part of the galaxy where kinematic data is available and dark matter does not play a significant role.

6.4.2 Dynamical models

As in Chapters 4 and 5 (Nowak et al., 2008, 2007) we use an axisymmetric code (Gebhardt et al., 2000a, 2003; Richstone & Tremaine, 1988; Thomas et al., 2004) to determine the mass of the SMBH in NGC 3368. This method has been successfully tested on the maser galaxy NGC 4258 in Siopis et al. (2009), who obtained the same mass for the black hole as determined from maser emission. Using an axisymmetric code for a barred and therefore obviously non-axisymmetric galaxy might be debatable, but can be justified in this case, as we only model the central part of the galaxy. Near the SMBH the potential is intrinsically spherical and strong non-axisymmetries are unlikely. Further out, however, non-axisymmetric structures could appear in projection over the SINFONI field of view, e.g. bar orbits that cross the centre. However, there is very little evidence for the presence of such orbits (see Section 6.4.4).

We use only the SINFONI data for the modelling. The four quadrants are modelled separately in order to assess the influence of deviations from axisymmetry. These four independent measurements of the SMBH mass should agree within the observational errors, if the data are axisymmetric. If they are not axisymmetric, the systematic differences from quadrant to quadrant provide an estimate for the systematic errors introduced by assuming axial symmetry.

Since the observed ellipticity of the classical bulge is affected by strong dust lanes, the ellipticity is slightly uncertain (see Section 6.2.2). Thus, in order to test the effects of varying the classical-bulge ellipticity, we run models with the bulge ellipticity set to a constant value of 0.0 (our best guess for the true, unextincted ellipticity) and 0.1 (the observed, but dust-affected ellipticity).

In order to find out whether the results depend on the assumed inclination of the galaxy, we run models for four different inclinations around the most likely value of 53°.

The SINFONI observations mainly cover the classical bulge region, so the disc Υ can only be weakly constrained. It could be better constrained if we included other kinematic data extending further out, but this has several disadvantages. The inconsistencies between the SINFONI and the optical measurements from the literature mean that the models could have difficulties fitting the different kinematic data sets reasonably well at the same time. In addition, the non-axisymmetries due to the two bars would be more noticeable at large radii, and a dark halo would become important.

6.4.3 Results

The results for $i = 53°$ and $\epsilon = 0.0$ are shown in Fig. 6.19 ($\Delta\chi^2$ as a function of $M_\bullet$ and the Υ of one component, marginalised over the other component's Υ), and for all inclinations and ellipticities in Fig. 6.20 (total χ^2 as a function of one of the three parameters $M_\bullet$, $\Upsilon_{\mathrm{bulge}}$, Υ_{disc}). The best-fitting values with 3 σ errors are listed in Table 6.4 for all four inclinations. The resulting best-fitting values for $M_\bullet$, $\Upsilon_{\mathrm{bulge}}$ and Υ_{disc} agree very well within $\lesssim 2\,\sigma$ between the four quadrants. For $i = 53°$ the mean black hole mass for the four quadrants is $\langle M_\bullet \rangle = 7.5 \times 10^6$ M$_\odot$ (rms($M_\bullet$) $= 1.1 \times 10^6$ M$_\odot$).

The resulting black hole mass does not depend much on the particular choice of the mass-to-light ratio of the disc Υ_{disc}, but decreases for increasing $\Upsilon_{\mathrm{bulge}}$. As shown in Fig. 6.20, the results also do not change systematically with the inclination. This shows that the inclination cannot be constrained better by dynamical modelling than by a thorough analysis of photometric data, and that a very precise knowledge of the inclination is not necessary for dynamical modelling purposes. Uncertainties in the ellipticity seem to have a larger effect than uncertainties in the inclination. While the resulting $M_\bullet$, Υ_{disc} and $\Upsilon_{\mathrm{bulge}}$ for $\epsilon = 0.1$ agree quite well with the $\epsilon = 0.0$ results, the χ^2 curves (dashed lines in Fig. 6.20) show a larger scatter in $\Upsilon_{\mathrm{bulge}}$.

The fit of the best model at $i = 53°$ and $\epsilon = 0.0$ to v, σ, h_3 and h_4 along the major axis is shown in Fig. 6.21 for all quadrants. The corresponding best fit without black hole would be very similar, which is why we choose to plot the differences in χ^2 instead (see below and Fig. 6.22).

The asymmetry of the data is reflected in the error bars. For quadrants 1 and 4, which are the quadrants with the higher velocity dispersion, the error bars are much larger than for quadrants 2 and 3. In 2 and 3 the no-black hole-solution is excluded by $\gtrsim 5\,\sigma$, whereas for 1 and 4 it is only excluded by $\sim 3.6\sigma$. The error contours are in general very large, allowing a large range of black hole masses.

In order to illustrate the significance of the result and where the influence of the black hole is largest, Fig. 6.22 shows the χ^2 difference between the best-fitting model without a black hole and the best-fitting model with a black hole for all LOSVDs in all four quadrants. As for NGC 4486a (Chapter 4) and Fornax A (Chapter 5), the largest signature seems to come mainly from the bins within about 2-3 spheres of influence, and the fit with black hole is generally better in about 68% of all bins. $\Upsilon_{\mathrm{bulge}}$ is large for the best models without black hole, which can worsen the fit in the

6.4. DYNAMICAL MODELLING OF NGC 3368

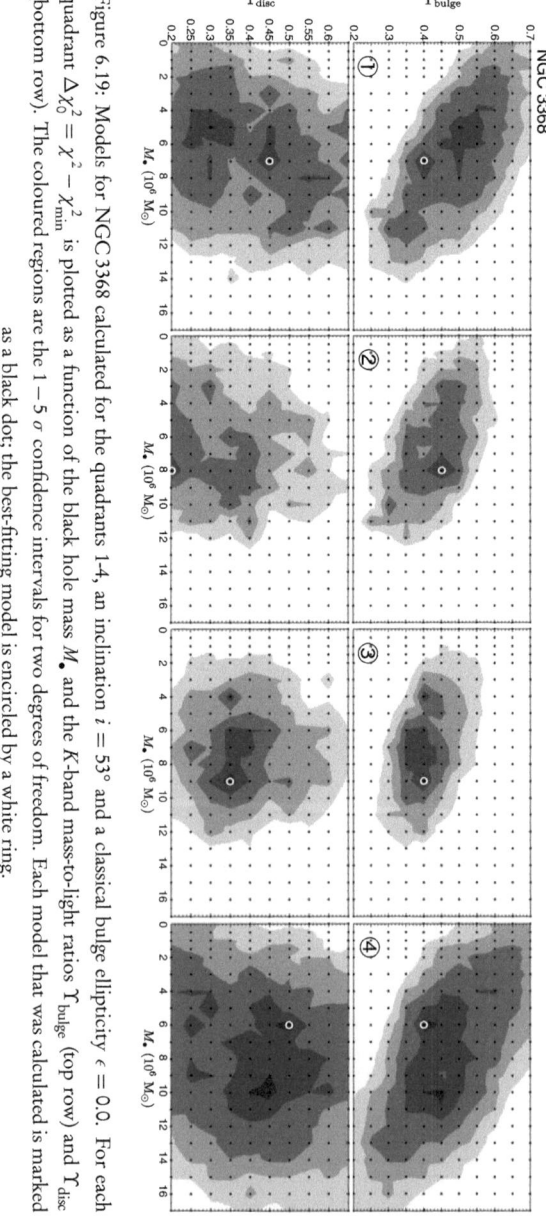

Figure 6.19: Models for NGC 3368 calculated for the quadrants 1-4, an inclination $i = 53°$ and a classical bulge ellipticity $\epsilon = 0.0$. For each quadrant $\Delta \chi^2_Q = \chi^2 - \chi^2_{min}$ is plotted as a function of the black hole mass $M_\bullet$ and the K-band mass-to-light ratios Υ_{bulge} (top row) and Υ_{disc} (bottom row). The coloured regions are the $1 - 5\,\sigma$ confidence intervals for two degrees of freedom. Each model that was calculated is marked as a black dot; the best-fitting model is encircled by a white ring.

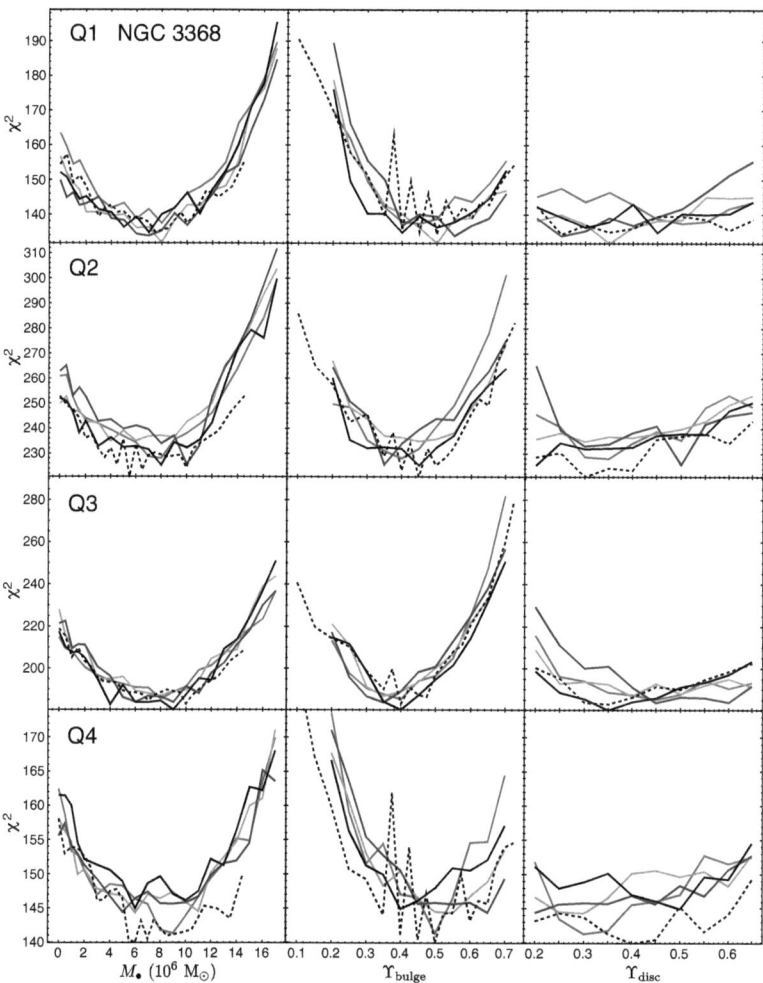

Figure 6.20: χ^2 as a function of $M_\bullet$ (left column, marginalised over Υ_{bulge} and Υ_{disc}), Υ_{bulge} (middle column, marginalised over $M_\bullet$ and Υ_{disc}) and Υ_{disc} (right column, marginalised over $M_\bullet$ and Υ_{bulge}) for NGC 3368. Different inclinations are indicated by different colours (red: $i = 52°$, black: $i = 53°$, green: $i = 54°$, blue: $i = 55°$). Models calculated for an ellipticity of the classical bulge of $\epsilon = 0.0$ are plotted as solid lines, models with $\epsilon = 0.1$ are plotted with dashed lines.

6.4. DYNAMICAL MODELLING OF NGC 3368

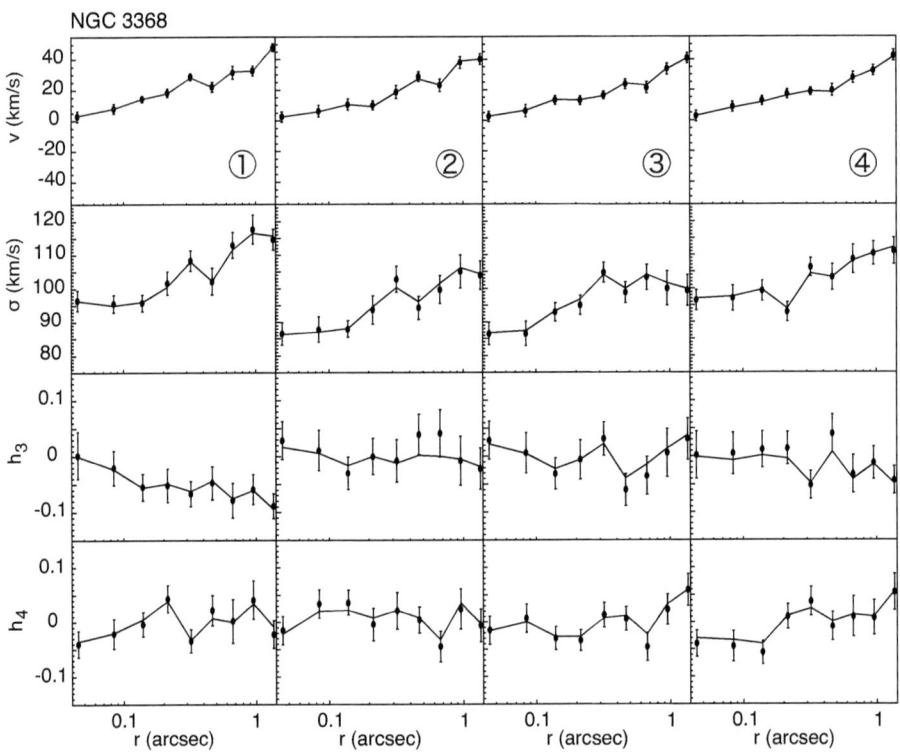

Figure 6.21: Fit (solid line) of the best models in quadrants 1 to 4 to the major axis kinematics (points) of NGC 3368.

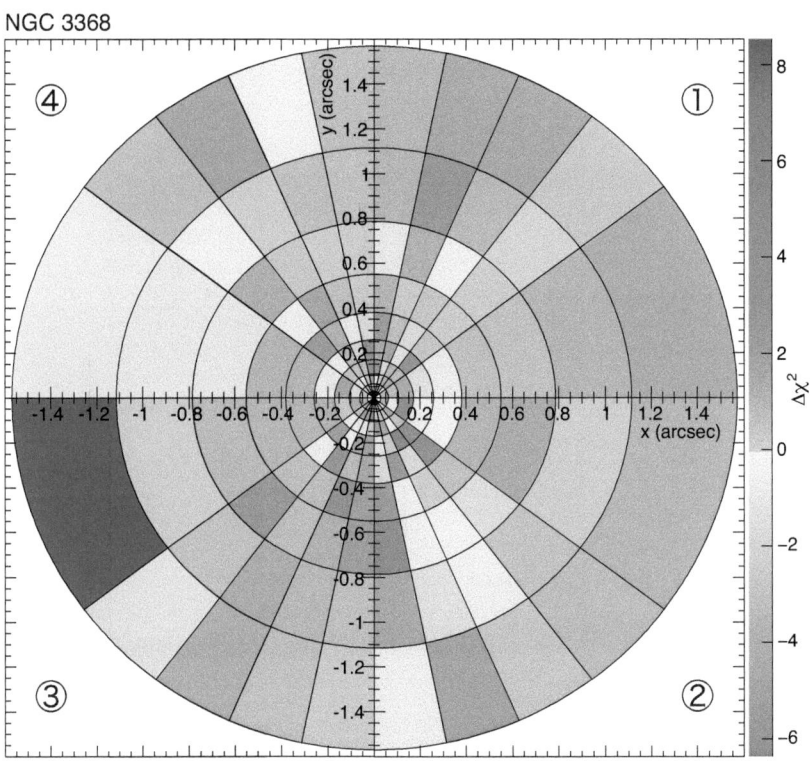

Figure 6.22: χ^2 difference between the best-fitting model without black hole and the best-fitting model with black hole ($\Delta\chi^2 = \sum_i \Delta\chi_i^2 = \sum_{i=1}^{21}(\chi_{i,\mathrm{noBH}}^2 - \chi_{i,\mathrm{BH}}^2)$ over all 21 velocity bins) for all LOSVDs of the four quadrants of NGC 3368. Bins where the model with black hole fits the LOSVD better are plotted in green, the others in orange.

6.4. DYNAMICAL MODELLING OF NGC 3368

Table 6.4: Resulting black hole masses and K-band mass-to-light ratios Υ_{bulge} and Υ_{disc} of NGC 3368 for all modelled quadrants (Q) and inclinations i. The lower and upper 3 σ limits are given in brackets. The total χ^2 of the best model with black hole and the χ^2 difference between the best model without black hole and the best model with black hole are given in the last two columns.

i	Q	$M_\bullet$ ($10^6\,M_\odot$)	Υ_{bulge}	Υ_{disc}	χ^2_{min}	$\Delta\chi^2_{\text{noBH}-\text{BH}}$
52°	1	8.0 (3.0, 11.0)	0.45 (0.30, 0.60)	0.50 (0.20, 0.65)	137.353	26.093
	2	8.0 (4.0, 10.0)	0.40 (0.30, 0.45)	0.35 (0.30, 0.50)	228.051	32.828
	3	8.0 (3.0, 10.0)	0.40 (0.25, 0.50)	0.45 (0.25, 0.65)	185.363	29.471
	4	9.0 (1.0, 13.0)	0.50 (0.30, 0.55)	0.30 (0.25, 0.65)	141.261	21.207
53°	1	7.0 (1.5, 11.0)	0.40 (0.30, 0.65)	0.45 (0.20, 0.65)	134.681	17.285
	2	8.0 (3.0, 10.0)	0.45 (0.25, 0.50)	0.20 (0.20, 0.40)	225.221	26.828
	3	9.0 (4.0, 10.0)	0.40 (0.35, 0.45)	0.35 (0.25, 0.50)	180.291	36.955
	4	6.0 (1.5, 14.0)	0.40 (0.25, 0.65)	0.50 (0.20, 0.65)	144.920	16.538
54°	1	8.0 (2.0, 11.0)	0.50 (0.35, 0.60)	0.35 (0.20, 0.50)	131.651	25.002
	2	4.0 (1.0, 11.0)	0.45 (0.30, 0.55)	0.30 (0.20, 0.55)	234.792	15.106
	3	8.0 (3.0, 11.0)	0.35 (0.30, 0.45)	0.40 (0.25, 0.65)	186.925	41.063
	4	6.0 (1.0, 14.0)	0.55 (0.30, 0.70)	0.30 (0.20, 0.65)	144.291	13.807
55°	1	7.0 (0.5, 12.0)	0.55 (0.40, 0.65)	0.25 (0.20, 0.50)	133.730	16.099
	2	10.0 (8.0, 11.0)	0.35 (0.35, 0.40)	0.50 (0.30, 0.50)	225.629	37.159
	3	9.0 (4.0, 11.0)	0.35 (0.25, 0.40)	0.45 (0.40, 0.65)	183.933	37.418
	4	5.0 (0.0, 15.0)	0.65 (0.30, 0.70)	0.20 (0.20, 0.65)	144.406	11.170

outer data regions. Therefore it is not surprising that improvements of the fit appear at all radii. As in the case of Fornax A (Chapter 5), the largest χ^2 differences appear in the high-velocity tails of the LOSVDs. The total $\Delta\chi^2$, summed over all LOSVDs, between the best-fitting model without a black hole and the best-fitting model with a black hole is given in the last column of Table 6.4 for each quadrant.

6.4.4 Discussion

Upper limits for $M_\bullet$ in NGC 3368 have been measured from central emission-line widths by Sarzi et al. (2002) and Beifiori et al. (2009), who obtain $2.7 \times 10^7\,M_\odot$ (stellar potential included) and $4.8 \times 10^7\,M_\odot$ (stellar potential not included), respectively. Based on the $M_\bullet$-σ relation of Tremaine et al. (2002), a black hole with a mass between 8×10^6 and $2.5 \times 10^7\,M_\odot$ would have been expected, depending on which σ measurement is used. From the relation between $M_\bullet$ and K-band luminosity (Marconi & Hunt, 2003) we would have expected a very high black hole mass of $9.2 \times 10^7\,M_\odot$ if it correlates with the total (photometric) bulge luminosity ($M_K = -23.42$), or a very small mass of only 1.5×10^6 $M_\odot$ if it correlates with the classical bulge luminosity ($M_K = -19.48$).

If the best-fitting mass for the black hole of $M_\bullet = 7.5 \times 10^6$ $M_\odot$ were entirely composed of stars, the mass-to-light ratio $\langle \Upsilon_{\text{bulge}} \rangle \approx 0.41$ would increase to 0.95. This would be typical for an older stellar population ($\sim 4-7$ Gyr for a Salpeter IMF and $\sim 10-11$ Gyr for a Kroupa IMF at solar metallicity, using the models of Maraston 1998, 2005). However, this would strongly conflict with Sarzi et al. (2005), who find that an 1 Gyr old population dominates, with some contributions from older and younger populations, resulting in a mean age of 3 Gyr.

As mentioned in Section 6.4.1, bar orbits crossing the centre could in principle produce deviations from axisymmetry. Non-axisymmetric structures such as a prolate central structure can be recognised in the kinematics as a low-σ, high-h_4 region if seen edge-on, or as a high-σ, low-h_4 region if seen face-on (Thomas et al., 2007). This could also bias the reconstructed masses. However, this is only valid for N-body ellipticals with a central prolate structure. Simulations of bars by Bureau & Athanassoula (2005) resulted in variable σ and h_4, depending on the projection of the bar. However, these variations were always symmetric with respect to the bar. Thus we would expect symmetry between quadrants where the bar appears. This is not the case, therefore it is unlikely that bar orbits in projection disturb the central kinematics significantly.

Dust could in principle influence the kinematics and produce distortions or asymmetries, though no clear correlation with the dust distribution could be found in Section 6.3.2. According to Baes et al. (2003), however, dust attenuation should not affect moderately inclined galaxies significantly.

The mass of the hot H_2 in the clouds and in the total field of view was estimated via $M_{H_2} = 5.0875 \times 10^{13} D^2 I_{1-0S(1)}$ (Rodríguez-Ardila et al., 2005) and the total mass of the cold gas was estimated via $\log(L_{1-0S(1)}/M_{H_2}^{\text{cold}}) = -3.6 \pm 0.32$ (Mueller Sánchez et al., 2006), where D is the distance of the galaxy in Mpc, I is the H_2 $(1-0)$ flux and L is the H_2 $(1-0)$ luminosity. They are listed in Table 6.2. Note that the latter conversion has large uncertainties and that the ratio $M_{H_2}^{\text{hot}} : M_{H_2}^{\text{cold}}$ spans at least two orders of magnitude ($\sim 10^{-7} - 10^{-5}$). It depends on the far-infrared colour $f_\nu(60\mu m)/f_\nu(100\mu m)$ (Dale et al., 2005), which may help to constrain the ratio a little bit. With a far-IR colour of 0.35 (Sakamoto et al., 1999) the ratio would be approximately $10^{-8\pm1}$. This is consistent with the ratio of 4×10^{-7} obtained using the Mueller Sánchez et al. (2006) conversion. A more precise way to constrain the cold H_2 masses is by more direct measurements of CO ($J = 1-0$) emission in the millimeter range. Helfer et al. (2003) measured a peak molecular surface density (i.e. the peak from the 6″ beam size) of 815 $M_\odot$ pc^{-2}. Over a 3″ aperture a total gas mass of approximately 1.5×10^7 $M_\odot$ would then be expected. Sakamoto et al. (1999) ($\sim 3″$ beam size) report a molecular mass of 4×10^8 $M_\odot$ (2.67×10^8 $M_\odot$ when using the CO-to-H_2 conversion factor of

Helfer et al. 2003) within a 15″ diameter aperture. Fig. 6.23 shows the measured gas mass distribution compared to (1) the dynamical mass distribution and (2) the gas masses obtained from CO ($J = 1 - 0$) measurements. We used figure 2 and equation 1 of Sakamoto et al. (1999) with the CO-to-H_2 conversion factor of Helfer et al. (2003) to estimate the gas mass distribution at smaller radii. Within a 3″ diameter aperture the gas masses from Sakamoto et al. (1999) and Helfer et al. (2003) agree very well. This shows that we overestimated the cold gas mass traced by hot H_2 by a factor of ~ 11. When comparing the gas mass profile (calibrated to match the gas mass at $r = 1.5″$ derived from CO) with the dynamical mass, we find that within a 3″ aperture the gas is approximately 5% of the dynamical mass. At smaller radii this fraction is even lower, as the gas mass profile is steeper than the dynamical mass profile. Therefore we do not expect that the gas has a noticeable influence on the stellar kinematics and, although the gas mass in quadrants 1 and 4 is larger than in quadrants 2 and 3, it is unlikely the cause for the asymmetry in the stellar kinematics; it might, however, be related to whatever is responsible for the latter.

Concerning the results of the black hole mass measurement the gas mass has no significant influence. As the gas mass distribution has a similar radial profile as the stellar mass, Υ_{bulge} likely includes the gas mass, such that the true stellar mass-to-light ratio is slightly lower than Υ_{bulge}. But this is no serious problem for the modelling. However, the fact that the gas is not evenly distributed and the clouds already have a mass of order 10^6 $M_\odot$ each, can weaken the evidence for the presence of a black hole, as it would imply that the centre is slightly out of equilibrium.

6.5 Dynamical modelling of NGC 3489

6.5.1 Construction of the luminosity profile for modelling

Given the apparent similarity of NGC 3489's inner structure to that of NGC 3368 (modulo the presence of a secondary bar in NGC 3368), including the strong isophotal twist created by the bar in NGC 3489, we followed a similar strategy for constructing the luminosity profiles. That is, we divide the galaxy into separate disc (which includes the discy pseudobulge) and central classical bulge components, with the disc treated as having a constant observed ellipticity of 0.41. The disc surface brightness profile is an azimuthal average with fixed ellipticity down to $r = 4.9″$, with the profile at smaller radii being the extrapolated inner-exponential fit from Fig. 6.10.

The classical bulge profile is the result of a free-ellipse fit to the inner-disc-subtracted WFPC2 F814W image. The latter was created by generating a model disc with ellipticity = 0.41 and profile

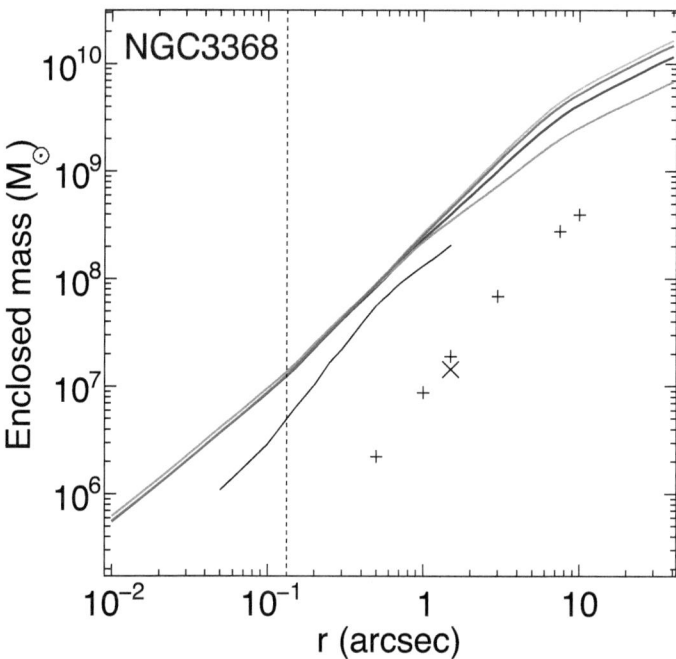

Figure 6.23: Enclosed mass as a function of radius in NGC 3368 for molecular mass (black line) and stellar dynamical mass. For the latter, we plot values estimated from our modelling of the individual quadrants (Q1: red, Q2: green, Q3: blue, Q4: cyan). The molecular gas mass is estimated from the hot H_2 emission in the SINFONI data, converted into a cold gas mass using Mueller Sánchez et al. (2006). The cold gas mass distribution directly measured from CO emission by Sakamoto et al. (1999) is marked by crosses and the cold gas mass from BIMA SONG (Helfer et al., 2003) is marked as an '×'. The cold gas mass derived using the hot H_2 conversion is clearly overestimated compared to the mass derived from CO. The vertical dashed line marks approximately the radius of the sphere of influence, defined as the radius that encloses a stellar mass of $M_* = 2M_\bullet$.

6.5. DYNAMICAL MODELLING OF NGC 3489

matching the exponential part of the fit in Fig. 6.10 (scale length = 4.9″), and then subtracting it from the dust-corrected WFPC2 image.

The deprojection was done in the same way as for NGC 3368 for bulge and disc component separately, but only for one inclination ($i = 55°$).

6.5.2 Dynamical models

NGC 3489 has only a weak large-scale bar and no nuclear bar. The measured kinematics and line indices are largely symmetric apart from the asymmetry in v. Thus non-axisymmetries are not expected to play a role as big as in NGC 3368. We first use only SINFONI data to model all four quadrants separately. However, we expect that, as for NGC 3368, due to the small field of view of the SINFONI data it will be difficult to constrain Υ_{disc}, as the data cover only that part of the galaxy where the classical bulge dominates. Thus we try to constrain Υ_{disc} beforehand by modelling SAURON and OASIS data alone. As the SAURON data have a large field of view including the bar, we use just the inner 10″ for that purpose. Finally, we model the combined SINFONI plus OASIS and/or SAURON data set.

We do not calculate models for different inclinations, as the inclination is well determined from the photometry ($i = 55°$). As shown for NGC 3368 in the previous section, the inclination cannot be constrained better via dynamical modelling and the differences within a small inclination range of a few degrees are small (see Tab. 6.4 and Fig. 6.20).

6.5.3 Results

In order to constrain Υ we first calculate models using symmetrised SAURON and OASIS kinematics separately. We only calculate models with $M_\bullet = 0$ for the SAURON data, but vary $M_\bullet$ between 0 and 1.3×10^7 $M_\odot$ for the OASIS data. Fig. 6.24a shows $\Delta \chi^2$ as a function of Υ_{bulge} and Υ_{disc} for the SAURON models. Υ_{disc} is well constrained, but in Υ_{bulge} a very large range between ~ 0 and ~ 0.68 is possible. This is due to the fact that the classical bulge is only just resolved with SAURON ($R_e^{CB} = 1.3″$, SAURON spatial resolution = 1.1″). The best-fitting model has $\Upsilon_{bulge} = 0.28$ and $\Upsilon_{disc} = 0.44$. For the OASIS models (Fig. 6.24b) the resulting Υ_{disc} is higher, which could be a result of the higher σ of the OASIS data compared to SAURON. Due to the higher spatial resolution (0.69″), Υ_{bulge} is better constrained. The best-fitting model has $\Upsilon_{bulge} = 0.6$ and $\Upsilon_{disc} = 0.36$. It is not possible to constrain $M_\bullet$ with the OASIS data alone (see Fig. 6.27).

CHAPTER 6. THE PSEUDOBULGE GALAXIES NGC 3368 AND NGC 3489

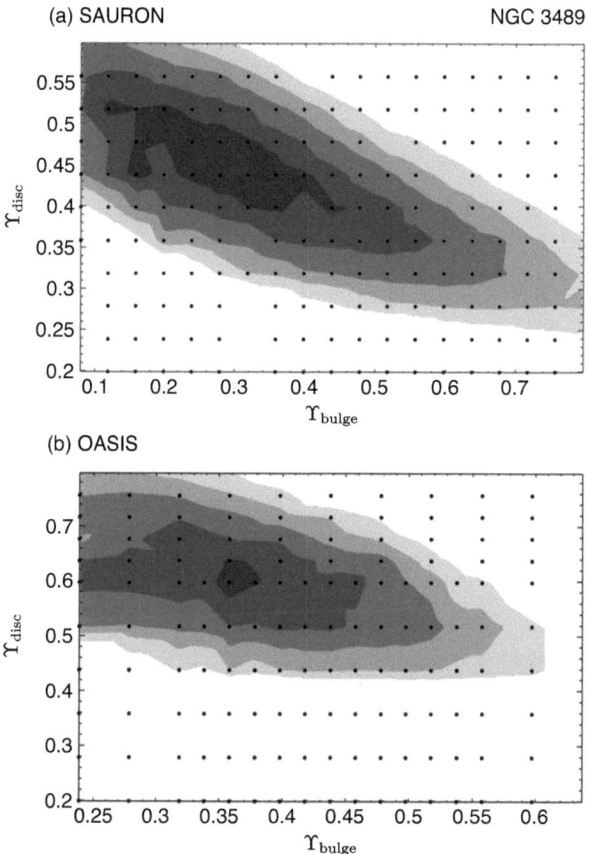

Figure 6.24: $\Delta\chi_0^2 = \chi^2 - \chi_{min}^2$ as a function of Υ_{disc} and Υ_{bulge} for (a) the symmetrised SAURON data and (b) the symmetrised OASIS data of NGC 3489. The coloured regions are the $1-5\,\sigma$ confidence intervals for two degrees of freedom. Each calculated model is marked as a black dot.

6.5. DYNAMICAL MODELLING OF NGC 3489

To derive the mass of the SMBH we first use the SINFONI kinematics alone to model the four quadrants separately. We chose a few values for Υ_{disc} around 0.44. Fig. 6.25 shows the resulting $\Delta\chi^2$ contours. The mean black hole mass for the four quadrants is $\langle M_\bullet \rangle = 4.25 \times 10^6$ M$_\odot$ (RMS($M_\bullet$)= 2.05×10^6 M$_\odot$). $M_\bullet$ clearly anticorrelates with Υ_{bulge}, but as in the case of NGC 3368 it does not depend on the specific choice of Υ_{disc}. The mean black hole mass for any fixed Υ_{disc} is consistent with the result for any other Υ_{disc} within 1 σ. The results with the corresponding 3 σ errors are given in Table 6.5 for all quadrants.

The error contours are large, such that a wide range of black hole masses is allowed. A solution without black hole is allowed in three quadrants within $2-4$ σ and in one quadrant even within 1 σ. Thus there is no evidence for the presence of a SMBH in one quadrant, and only weak evidence in the others, when modelling the SINFONI data alone. The fit of the best model in each quadrant to v, σ, h_3 and h_4 along the major axis is shown in Fig. 6.26a.

The resulting $M_\bullet$, Υ_{bulge} and Υ_{disc} of the four quadrants agree with each other within < 2 σ, and as there are also no obvious strong inconsistencies between the kinematics of the quadrants, we fold the LOSVDs of the four quadrants (the LOSVDs of quadrants 1 and 4 were also flipped, such that v and h_3 change sign). For the folded data we find a best-fitting black hole mass of $M_\bullet = 5.0 \times 10^6$ M$_\odot$ at $\Upsilon_{\text{bulge}} = 0.56$. This is in good agreement with the results of the individual quadrants. A solution without black hole is allowed within 3 σ, thus as a conservative result we can only give a 3 σ upper limit of 1.3×10^7 M$_\odot$ for the SMBH in NGC 3489, when using just the SINFONI data.

The non-dependence of $M_\bullet$ on Υ_{disc} can be explained by the very small field of view of the SINFONI data, which covers only the very central part of the galaxy, dominated by the classical bulge. This might also explain the relatively weak detection of a SMBH in NGC 3489 despite the high quality data. It therefore seems reasonable to include kinematics at larger radii, like the SAURON or the OASIS kinematics, as these data sets cover a large fraction of the disc and therefore are able to constrain Υ_{disc} very well, as shown above. We should keep in mind however, that the SAURON and OASIS velocity dispersions do not fully agree with each other, are larger than the SINFONI dispersion and show some deviations from axisymmetry, which might possibly be due to the strong dust features. In order to determine how strong these differences affect the result of the modelling we do three sets of models: the first one with SINFONI and OASIS data (using OASIS data between $r = 0.5''$ and $4''$), the second one with SINFONI and SAURON data (using SAURON data between $r = 1''$ and $10''$) and the third one with all three data sets (with OASIS data between $r = 0.5''$ and $4''$ and SAURON data between $4''$ and $10''$).

Fig. 6.27 shows the resulting $\Delta\chi^2$ profiles for all data combinations we used. It is clear that the mass of the black hole can be much better constrained when including SAURON and/or OASIS

CHAPTER 6. THE PSEUDOBULGE GALAXIES NGC 3368 AND NGC 3489

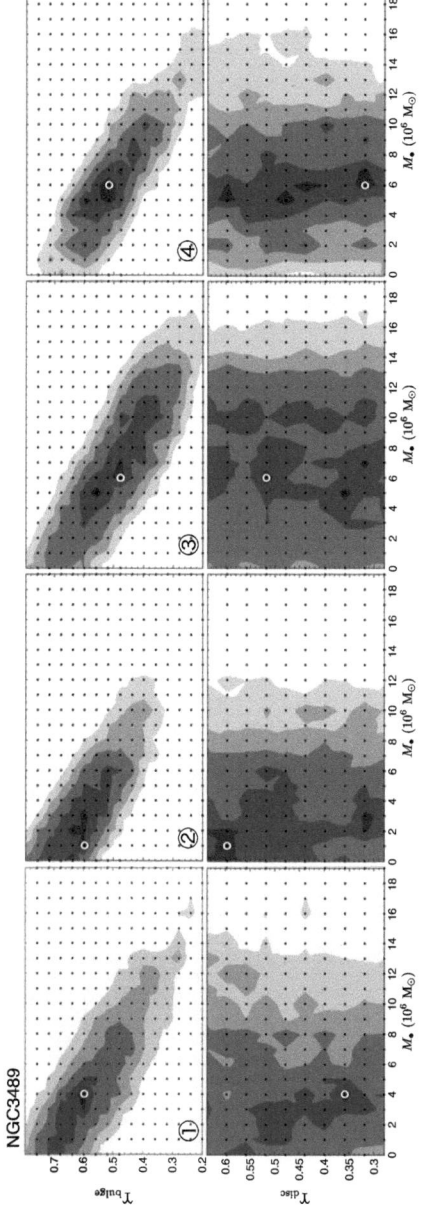

Figure 6.25: Same as Fig. 6.19 for NGC 3489. The models were calculated with SINFONI data only for the quadrants 1-4 and an inclination $i = 55°$.

6.5. DYNAMICAL MODELLING OF NGC 3489

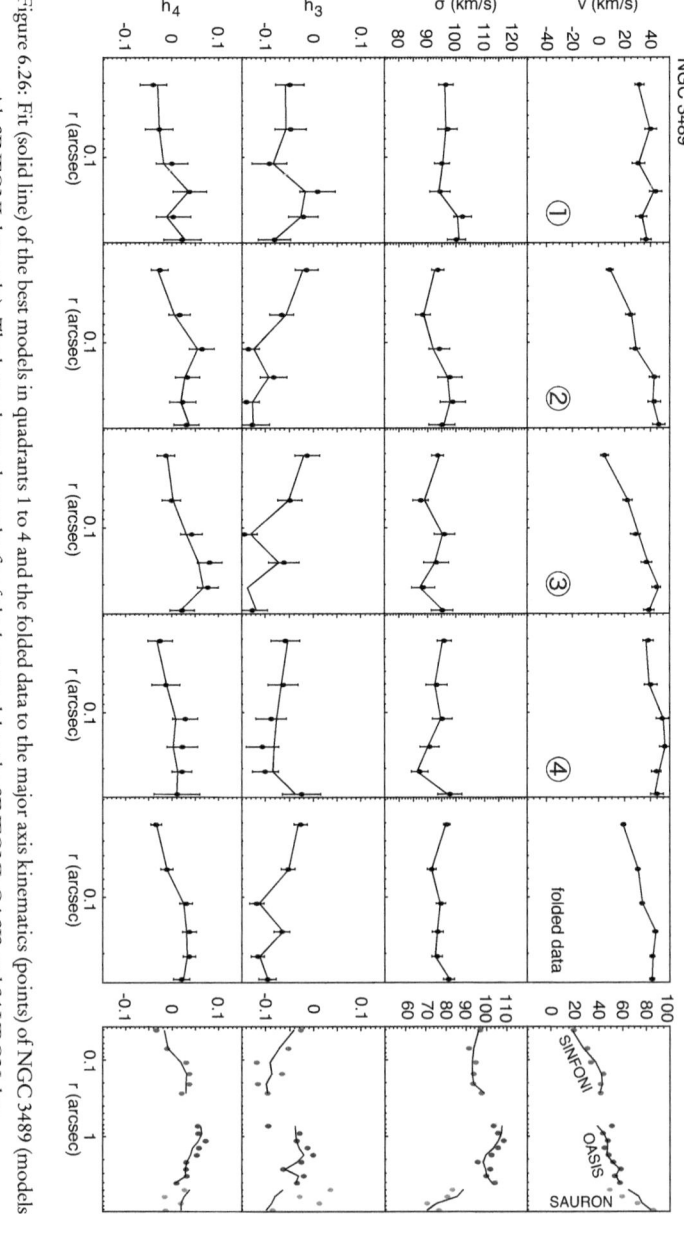

Figure 6.26: Fit (solid line) of the best models in quadrants 1 to 4 and the folded data to the major axis kinematics (points) of NGC 3489 (models with SINFONI data only). The last column shows the fit of the best model to the SINFONI, OASIS and SAURON data.

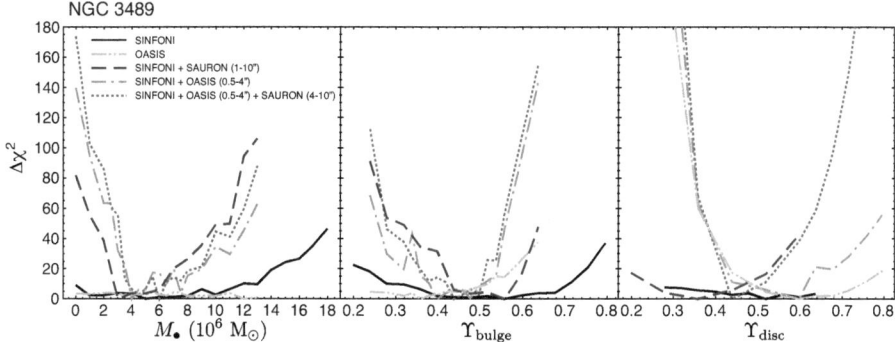

Figure 6.27: $\Delta\chi_0^2 = \chi^2 - \chi_{min}^2$ as a function of $M_\bullet$ (left column, marginalised over Υ_{bulge} and Υ_{disc}), Υ_{bulge} (middle column, marginalised over $M_\bullet$ and Υ_{disc}) and Υ_{disc} (right column, marginalised over $M_\bullet$ and Υ_{bulge}) for NGC 3489. The colours indicate the data sets used for the modelling (black: SINFONI, grey: OASIS, blue: SINFONI+SAURON, green: SINFONI+OASIS, red: SINFONI+OASIS+SAURON).

data. The constraints on Υ_{bulge} and Υ_{disc} are also much stronger in these cases. Using SAURON data in addition to SINFONI and OASIS does not seem to improve the measurement of $M_\bullet$ and Υ_{bulge}. The scatter in the $\Delta\chi^2$ profiles is quite large for the models of the combined data sets, despite the good quality and high S/N of the individual data sets and despite the comparatively small scatter in the models of individual data sets. The uncertainties of the SMBH mass measurement therefore do not seem to be dominated by statistical errors, but instead by systematics. Systematic errors can be introduced e.g. due to the differences in the kinematics of the individual data sets. Systematic errors in the modelling (e.g. slightly different results for different quadrants) could add to the scatter as well, but are difficult to quantify. We measure the formal 1 σ errors (corresponding to $\Delta\chi^2 = 1$ for one degree of freedom) by fitting a third order polynomial to each curve in Fig. 6.27. The best values for $M_\bullet$, Υ_{bulge} and Υ_{disc} given in Table 6.6 refer to the minimum of the fit and the associated $\Delta\chi^2 \leqslant 1$ region. We then determine the systematic error, introduced by the differences between the data sets, from the scatter of the best fits for models with combined data sets. Thus when using all available data, we obtain a black hole mass of $M_\bullet = (6.00^{+0.56}_{-0.54}|_{stat} \pm 0.64|_{sys}) \times 10^6 \, M_\odot$, a bulge mass-to-light ratio $\Upsilon_{bulge} = 0.45 \pm 0.02|_{stat} \pm 0.03|_{sys}$ and a disc mass-to-light ratio $\Upsilon_{disc} = 0.47^{+0.01}_{-0.02}|_{stat} \pm 0.05|_{sys}$.

6.5.4 Discussion

No attempts have been made in the literature to measure the mass of the SMBH in NGC 3489. From the $M_\bullet$-σ relation of Tremaine et al. (2002) we would expect a mass between $M_\bullet = 5.2 \times 10^6$ M$_\odot$ for $\sigma_e = 88.9$ km s^{-1} derived from the SAURON data (Emsellem et al., 2004) and $M_\bullet = 9.2 \times 10^6$ M$_\odot$ for $\sigma_e = 102.5$ km s^{-1} derived from the OASIS data (McDermid et al., 2006). From the relation between $M_\bullet$ and K-band magnitude (Marconi & Hunt, 2003) we would expect a black hole mass of $M_\bullet = 1.92 \times 10^7$ M$_\odot$ if it correlates with the total bulge magnitude $M_{K,\text{total}} = -21.91$, or 4.94×10^6 M$_\odot$ if it correlates with the classical bulge magnitude $M_{K,\text{bulge}} = -20.60$ only.

Although with the OASIS data alone it is not possible to constrain the mass of the SMBH, the region covered by this data set ($\sim 0.5 - 4''$) seems to be crucial for the lower limit on $M_\bullet$, which is not possible to derive with the SINFONI data alone. This means that differences between models without black hole and models with black hole (say, $M_\bullet = 6 \times 10^6$ M$_\odot$) should not only appear within the sphere of influence, but also further outside. This is in agreement with the observations in NGC 3368, Fornax A and NGC 4486a (Sections 6.4.3, 5.5.2 and 4.6), where a general improvement of the fit even outside the sphere of influence, was observed.

The stellar population models of Maraston (1998, 2005) suggest an age of ~ 1 Gyr for the best-fitting $\Upsilon_{\text{bulge}} = 0.45 \approx \Upsilon_{\text{disc}}$ and a high metallicity [Z/H]~ 0.67, and an age of $\sim 2 - 3$ Gyr for a solar metallicity population (assuming a Salpeter IMF). This is in agreement with McDermid et al. (2006), who find both an age gradient (from $\sim 2 - 3$ Gyr in the outer regions to ~ 1 Gyr in the centre) and a metallicity gradient (from $\sim$solar in the outer regions to ~ 0.6 in the centre). It is also compatible with Sarzi et al. (2005), who found a mean age of ~ 3 Gyr assuming solar metallicity. If the best-fitting mass for the black hole of $M_\bullet = 6 \times 10^6$ M$_\odot$ were entirely composed of stars, the mass-to-light ratio of the bulge would increase to 0.78. This would be typical for an older stellar population (~ 5 Gyr for a high metallicity [Z/H]=0.67), and therefore conflict with the values found by Sarzi et al. (2005) and McDermid et al. (2006).

6.6 Summary and discussion

We analysed near-IR integral-field data for two barred galaxies that host both a pseudobulge and a classical bulge component. Both galaxies show fast and regular rotation and a σ-drop at the centre, which in the case of NGC 3368 is more pronounced and may have developed from gas, transported to the inner region by the bars and spiral arms. The kinematics of NGC 3368 – in particular the velocity dispersion – is asymmetric. The reasons for that could be dust or the non-axisymmetric

Table 6.5: Resulting black hole masses $M_\bullet$ and H-band mass-to-light ratios Υ_{bulge} and Υ_{disc} of the different quadrants of NGC 3489. The lower and upper 3 σ limits are given in brackets. The total χ^2 of the best model with black hole and the χ^2 difference between the best model without black hole and the best model with black hole are given in the last two columns.

Q	$M_\bullet$ ($10^6\,M_\odot$)	Υ_{bulge}	Υ_{disc}	χ^2_{min}	$\Delta\chi^2_{noBH-BH}$
1	4.0 (0.0, 8.0)	0.60 (0.44, 0.76)	0.36 (0.28, 0.64)	106.812	3.289
2	1.0 (0.0, 7.0)	0.60 (0.44, 0.80)	0.60 (0.28, 0.64)	100.554	0.301
3	6.0 (0.0, 13.0)	0.48 (0.28, 0.72)	0.52 (0.28, 0.64)	57.612	10.487
4	6.0 (1.0, 10.0)	0.52 (0.36, 0.64)	0.32 (0.28, 0.64)	80.077	16.786
folded	5.0 (0.0, 13.0)	0.56 (0.28, 0.72)	0.52 (0.28, 0.64)	47.877	9.060

Table 6.6: Resulting black hole masses $M_\bullet$ and H-band mass-to-light ratios Υ_{bulge} and Υ_{disc} of NGC 3489 for the folded SINFONI data alone (SINF) and in combination with SAURON (SAU) and OASIS (OA) kinematics. The lower and upper 1 σ limits (1 degree of freedom), determined by fitting a third order polynomial to the $\Delta\chi^2$ profiles of Fig. 6.27, are given in brackets. The total χ^2 of the best model with black hole and the χ^2 difference between the best model without black hole and the best model with black hole are given in the last two columns.

Data set	$M_\bullet$ ($10^6\,M_\odot$)	Υ_{bulge}	Υ_{disc}	χ^2_{min}	$\Delta\chi^2$
SINF	5.97 (3.64, 8.13)	0.53 (0.48, 0.58)	0.55 (0.48, 0.62)	106.812	3.289
SINF/SAU	4.56 (4.03, 5.12)	0.52 (0.50, 0.54)	0.36 (0.32, 0.40)	368.458	81.83
SINF/OA	5.81 (5.21, 6.46)	0.46 (0.44, 0.48)	0.52 (0.50, 0.55)	551.297	139.86
SINF/OA/SAU	6.00 (5.46, 6.56)	0.45 (0.43, 0.47)	0.47 (0.45, 0.48)	606.653	173.86

6.6. SUMMARY AND DISCUSSION

potential induced by the two bars. The gas distribution is also inhomogeneous, but as the total gas mass accounts for only $\lesssim 5\%$ of the dynamical mass, this has probably no significant influence on the stellar kinematics. There are two kinematically decoupled gas clouds located a few tens of parsecs north of the centre. Each cloud has a total mass of order $10^6 M_\odot$. The kinematics of NGC 3489 is very regular, with a slight asymmetry in the velocity field. All other kinematic parameters and the line indices are consistent with axisymmetry. No gas emission was found in NGC 3489. The near-IR line indices Na I and CO show a negative gradient in both galaxies, indicating an age and/or metallicity gradient.

We applied axisymmetric dynamical models to derive the SMBH masses in NGC 3368 and NGC 3489. In our models we assume that the galaxy potential can be decomposed into three components: the central black hole, an inner, classical bulge (with mass-to-light ratio Υ_{bulge}) and the outer disc (Υ_{disc}). The inclination of the models is fixed by the isophotes of the outer disc. For NGC 3368 we modelled the four quadrants of our IFU data independently and the resulting black hole masses and mass-to-light ratios agree very well. We find that $M_\bullet$ is largely independent of Υ_{disc} and anticorrelates with Υ_{bulge}. The average black hole mass for the four quadrants and an inclination $i = 53°$ is $\langle M_\bullet \rangle = 7.5 \times 10^6$ $M_\odot$ (RMS($M_\bullet$) = 1.1×10^6 $M_\odot$). A solution without a black hole is excluded by $\approx 4-5\ \sigma$. The error contours, however, cover a large range in $M_\bullet$. The largest uncertainty for $M_\bullet$ comes from the unknown Υ_{bulge}, and independent stellar population constraints would likely improve the results. The scatter from quadrant to quadrant is smaller than the uncertainty related to Υ_{bulge}, suggesting that the symmetry assumption plays a minor role for the uncertainty of $M_\bullet$. Our results do not significantly depend on the inclination (within the photometrically allowed inclination range).

For NGC 3489, modelling of the four SINFONI quadrants likewise gave consistent black hole masses and mass-to-light ratios. Similar to NGC 3368 the error contours cover a wide range of $M_\bullet$, the black hole mass is independent of Υ_{disc} and it clearly anticorrelates with Υ_{bulge}. Modelling the folded SINFONI data gives the same result as for the individual quadrants; thus, non-axisymmetries do not seem to play a role. When including OASIS and/or SAURON data, Υ_{bulge} and therefore also $M_\bullet$ could be much better constrained. Using all three data sets, we derived for NGC 3489 a SMBH mass of $M_\bullet = (6.00^{+0.56}_{-0.54}|_{stat} \pm 0.64|_{sys}) \times 10^6$ $M_\odot$ with a bulge mass-to-light ratio of $\Upsilon_{bulge} = 0.45 \pm 0.02|_{stat} \pm 0.03|_{sys}$ and a disc mass-to-light ratio $\Upsilon_{disc} = 0.47^{+0.01}_{-0.02}|_{stat} \pm 0.05|_{sys}$. A solution without a black hole is excluded with high significance. To derive a firm lower limit to $M_\bullet$, data between $\sim 0.5 - 4''$ seem to be crucial, in addition to the high-resolution SINFONI data in the centre. With OASIS data alone, no limits on $M_\bullet$ could be placed. There are some inconsistencies in the kinematics between the three data sets, which seems to be the main source of systematic errors.

In particular when modelling OASIS data alone, we get a higher $\Upsilon_{\rm disc}$ than if modelling SAURON data alone (because the inner σ is higher in the OASIS data than in the SAURON data).

The implications for the $M_\bullet$-σ relation and the $M_\bullet$-M_K relation are illustrated in Fig. 6.28. For NGC 3368 the mean $M_\bullet$ of the four quadrants and the RMS, and for NGC 3489 $M_\bullet$ from the combination of SINFONI, SAURON and OASIS data with its statistical 1 σ error is plotted against σ and M_K using the relations of Tremaine et al. (2002), Ferrarese & Ford (2005) and Marconi & Hunt (2003). All values for σ_e and $\sigma_{e/8}$ were measured using the effective radius of the total photometric bulge, as was done for all the galaxies contributing to the Tremaine et al. (2002) and Ferrarese & Ford (2005) relations. No attempt to determine σ_e for the classical components has therefore been made, but as we use luminosity-weighted measurements all values determined from high-resolution data represent mostly the classical bulge.

The agreement of NGC 3368 with the $M_\bullet$-σ relation largely depends on the value of σ which is used. The small $\sigma = 98.5$ km s^{-1} measured within the SINFONI field of view is in good agreement with the $M_\bullet$-σ relation. When combining the SINFONI σ with σ measurements of Whitmore, Schechter & Kirshner (1979), Héraudeau et al. (1999) and Vega Beltrán et al. (2001) a value of $\sigma_{e/8} = 104$ km s^{-1} is obtained (Erwin & Gadotti, in preparation). The velocity dispersions from the literature alone however (e.g. $\sigma_e = 117$ km s^{-1} estimated by Sarzi et al. 2002, $\sigma_e = 130.9$ km s^{-1} respectively $\sigma_{e/8} = 129.9$ km s^{-1} measured by Héraudeau et al. 1999, or $\sigma \approx 150$ km s^{-1} by Moiseev et al. 2004) are significantly larger than expected by this estimate and not or only marginally in agreement with the $M_\bullet$-σ relation. With a K-band magnitude of -23.42 for the total photometric bulge, NGC 3368 falls far (a factor of ~ 12) below the $M_\bullet$-M_K relation of Marconi & Hunt (2003). If we postulate that the SMBH only correlates with the magnitude of the classical bulge, the situation improves somewhat. With $M_K^{\rm CB} = -19.48$ NGC 3368 now lies a factor of ~ 5 above the $M_\bullet$-M_K relation; due to the large error contours in $M_\bullet$, it is thus in agreement with the relation within 3 σ errors.

For NGC 3489 the situation is similar. $M_\bullet$ is in excellent agreement with the $M_\bullet$-σ relation when using either the SINFONI mean $\sigma = 91.1$ km s^{-1} or the SAURON values $\sigma_e = 88.9$ km s^{-1} and $\sigma_{e/8} = 94$ km s^{-1}. It is still in reasonably good agreement with the relation when using the OASIS measurements ($\sigma_e = 102.5$ km s^{-1}, $\sigma_{e/8} = 108.9$ km s^{-1}) or when taking into account other σ measurements from the literature ($\sigma_{e/8} = 115$ km s^{-1} using Barth et al. 2002; Dalle Ore et al. 1991; Smith et al. 2000; Whitmore et al. 1979 and the SINFONI value). With a K-band magnitude of the total photometric bulge of $M_K^{\rm PB} = -21.91$ NGC 3489 also falls far below the $M_\bullet$-M_K relation of Marconi & Hunt (2003), but is in excellent agreement if the magnitude of the classical bulge component is considered ($M_K = -20.60$).

6.6. SUMMARY AND DISCUSSION

The large difference in the σ measurements makes it difficult to draw any firm conclusion with respect to the location of pseudobulges in the $M_\bullet$-σ relation, and at the same time illustrates that measurement errors in σ may play a larger role than one may have thought, in particular when dealing with small galaxy samples. NGC 3368 would fall far below the $M_\bullet$-σ relation when optical longslit kinematics alone are used. These discrepancies between the σ measurements might at least partly be due to dust, which affects the optical data much more than the near-IR data.

The K-band magnitudes on the other hand can be determined very accurately even for subcomponents of the galaxy. Taken at face value, both galaxies clearly do not follow the $M_\bullet$-M_K relation of Marconi & Hunt (2003) when considering the K-band magnitudes of the total photometric bulge, but are in better (NGC 3368) or even excellent (NGC 3489) agreement with it when considering the classical bulge magnitude only.

If we take into account that a stellar population becomes fainter when it ages passively (2.3 mag in K band for a solar metallicity population between 1 and 10 Gyr, based on the stellar population models of Maraston 1998, 2005), the pseudobulges would move toward the $M_\bullet$-M_K relation with time. Given the uncertainties on the age estimate, the exact size of the effect is unclear. Keeping in mind this caveat, this is in line with Greene et al. (2008), who conclude that pseudobulges follow the $M_\bullet$-σ relation, but not the $M_\bullet$-M_{bulge} relation, as well as with Gadotti & Kauffmann (2009), who find that pseudobulges follow only one of the two relations, if any. In order to strengthen our results, studies of a larger sample of pseudobulges similar in design are necessary.

Whether modelling single quadrants of obviously non-axisymmetric galaxies with an axisymmetric code is a good approximation and gives the correct black hole masses is certainly still an issue that remains to be resolved. The recently developed triaxial codes NMAGIC (de Lorenzi et al., 2007) and the code of van den Bosch et al. (2008) will have the potential to solve this issue in the future.

Acknowledgments

We would like to thank the Paranal Observatory Team for support during the observations. We are grateful to Harald Kuntschner and Mariya Lyubenova for providing us the code to measure near-IR line indices, and to Karl Gebhardt for providing the MPL code. Furthermore we thank Alexei Moiseev and Richard McDermid for providing us their 2D kinematics on NGC 3368 and NGC 3489. Finally we would also like to thank Maximilian Fabricius, Roland Jesseit and Erin Hicks for valuable discussions. This work was supported by the Cluster of Excellence: "Origin

CHAPTER 6. THE PSEUDOBULGE GALAXIES NGC 3368 AND NGC 3489

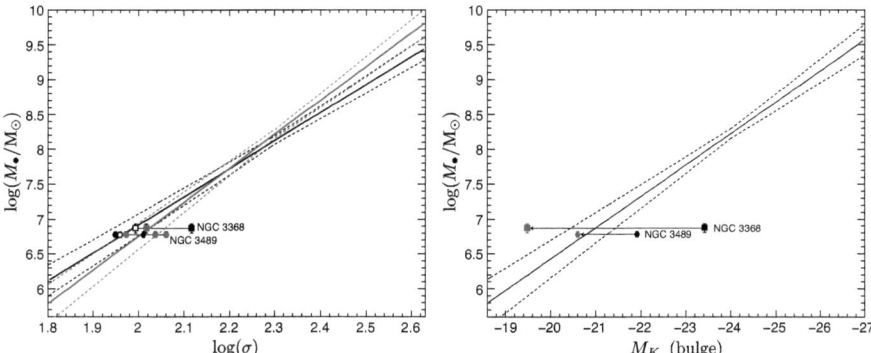

Figure 6.28: Left panel: Location of NGC 3368 and NGC 3489 with respect to the $M_\bullet$-σ relation (black: Tremaine et al. 2002, red: Ferrarese & Ford 2005). The velocity dispersion from our SINFONI measurements (open symbols) and values for σ_e (filled black symbols) and $\sigma_{e/8}$ (filled red symbols) derived from the literature are plotted for each galaxy. Right panel: Location of the two galaxies with respect to the $M_\bullet$-M_K relation of Marconi & Hunt (2003). The K-band magnitudes of the total photometric bulges are plotted as filled black symbols, the magnitudes of the classical bulge components as filled red symbols. In both panels the average $M_\bullet$ of the four quadrants and the RMS is plotted for NGC 3368, and $M_\bullet$ from the combination of SINFONI, SAURON and OASIS data with its statistical 1 σ error is plotted for NGC 3489.

6.6. SUMMARY AND DISCUSSION

and Structure of the Universe" and by the Priority Programme 1177 "Galaxy Evolution" of the Deutsche Forschungsgemeinschaft.

7
Concluding remarks

In the previous three chapters the masses of the central black hole in four galaxies were determined with stellar dynamics: the low-σ galaxy NGC 4486a, the merger remnant Fornax A and the two pseudobulge galaxies NGC 3368 and NGC 3489. Table 7.1 lists $M_\bullet$ with 1 σ errors and other properties according to Table 1.1 in Chapter 1. Figs. 7.2 and 7.3 show updated versions of the $M_\bullet$-σ and $M_\bullet$-L_K diagrams. The following paragraphs discuss the implications on the $M_\bullet$-bulge relations derived from the new measurements. The last part of this chapter gives an overview of the current status of the observations, the data reduction and the dynamical modelling of the remaining galaxies from Table 2.1.

Implications for the $M_\bullet$-σ relation

The location of the galaxies analysed in Chapters 4-6 with respect to the $M_\bullet$-σ relation of Tremaine et al. (2002) and Ferrarese & Ford (2005) is shown in Fig. 7.2a. The black hole masses of all four analysed galaxies (the low-σ classical bulge galaxy NGC 4486a, the two pseudobulges NGC 3368 and NGC 3489 and the merger remnant Fornax A) are in very good agreement with the $M_\bullet$-σ relation. The same is true for NGC 5102 ($\sigma \approx 60$ km s^{-1}), which is currently being modelled

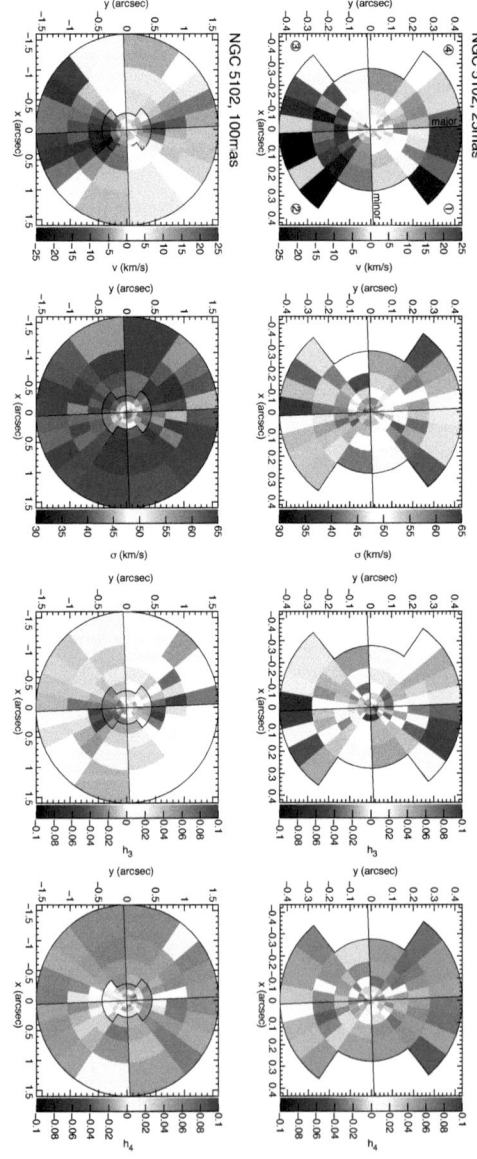

Figure 7.1: Stellar kinematic parameters v, σ, h_3 and h_4 of NGC 5102 for two different platescales: 25mas (upper row) and 100mas (bottom row). The major and the minor axis as well as the numeration of the different quadrants are indicated in the velocity field (upper left). The 25mas field of view is marked in the 100mas fields for a better orientation.

CHAPTER 7. CONCLUDING REMARKS

Table 7.1: Black hole masses measured in Chapters 4-6.

	NGC 1316	NGC 3368	NGC 3489	NGC 4486a
Type	E pec	SABab	SAB0	E
AGN	FRI, L	L2	T2/S2	
$M_\bullet$ ($M_\odot$)	1.3×10^8	7.5×10^6	6.00×10^6	1.26×10^7
($M_\bullet^{low}$, $M_\bullet^{high}$)	$(0.9, 1.7) \times 10^8$	$(6.4, 8.6) \times 10^6$	$(5.44, 6.56) \times 10^6$	$(0.85, 1.67) \times 10^7$
σ (km s^{-1})	^{e}226	^{8}104	^{e}88.9	110.5
D (Mpc)	18.6	^{t}10.4	^{t}12.1	16.0
PSF ($''$)	0.085	0.165	0.08	0.10
SoI ($''$)	0.24	0.12	0.11	0.11
M_K (total bulge)	-25.8	-23.42	-21.91	-22.11
M_K (class. bulge)	-25.8	-19.48	-20.60	-22.11

For NGC 3489 and NGC 4486a the black hole mass derived from the averaged LOSVDs and the 1 σ errors are given. For NGC 1316 and NGC 3368 the average black hole mass derived from the four individual quadrants together with the RMS error are given.

The K-band magnitude of NGC 4486a and $\sigma_{e/8}$ of NGC 3368 is taken from Erwin & Gadotti (in preparation)

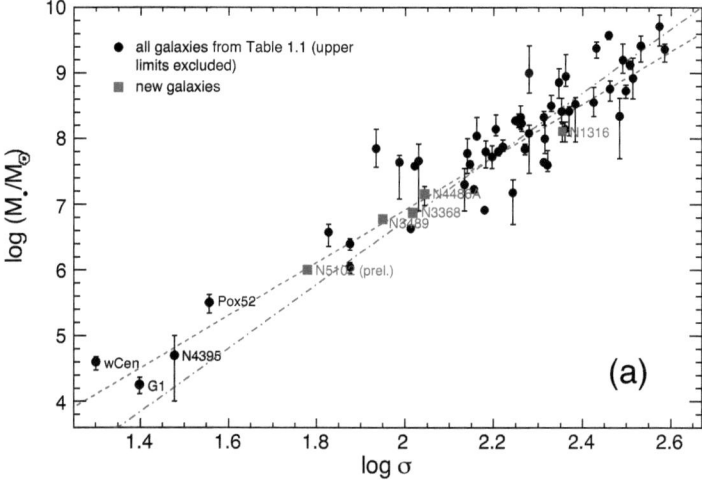

Figure 7.2: Updated $M_\bullet$-σ relation. The $M_\bullet$-σ relation of Tremaine et al. (2002) (dashed line) and Ferrarese & Ford (2005) (dot-dashed line) are plotted as a reference. (a) All galaxies from Table 1.1 (upper limits excluded) are plotted in black. The new $M_\bullet$ measurements from Table 7.1 are plotted in red. For NGC 5102 preliminary results (Thomas et al., in preparation) are shown.

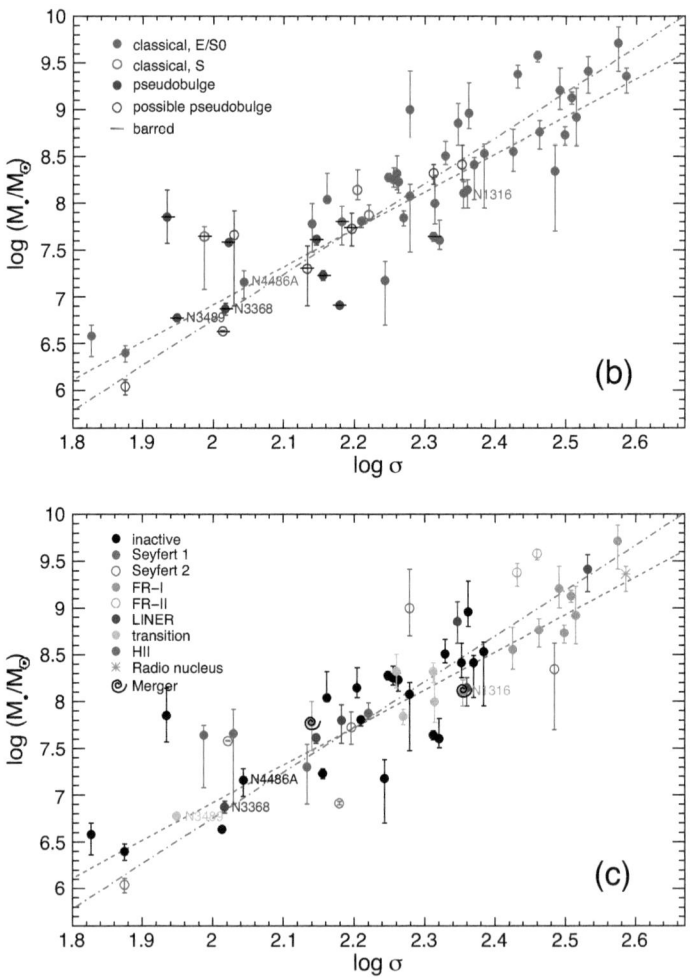

Figure 7.2: Continued. Updated $M_\bullet$-σ relation for relevant subsamples. The galaxies analysed in Chapters 4-6 are marked by their names. (b) Classical bulges (red) and pseudobulges (blue). Barred galaxies are marked by a black bar. (c) Inactive galaxies (black) and different types of AGN (coloured).

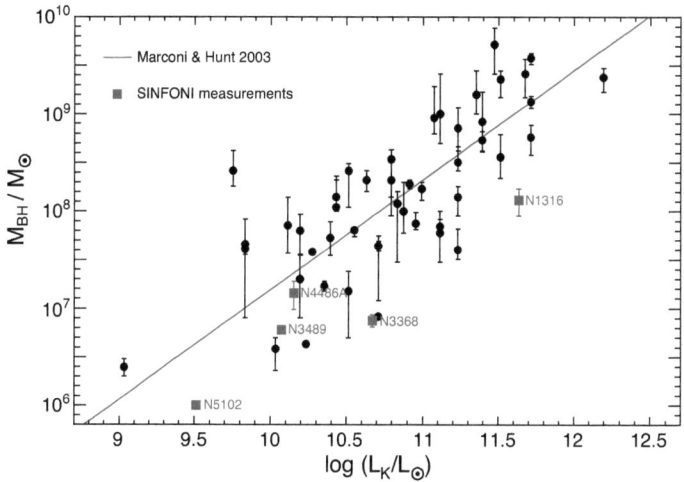

Figure 7.3: Updated $M_\bullet$-L_K relation. The $M_\bullet$-L_K relation of Marconi & Hunt (2003) (dashed line) is plotted as a reference. (a) All galaxies from Table 1.1 with measured L_K are plotted in black. The new $M_\bullet$ measurements from Table 7.1 are plotted in red. For NGC 5102 preliminary results are shown.

(Thomas et al., in preparation). The kinematics of NGC 5102 is shown in Fig. 7.1. The dynamical models indicate a black hole mass of about 1×10^6 M$_\odot$.

At the low-σ end ($\sigma \lesssim 130$ km s^{-1}) no indication for a deviation from the $M_\bullet$-σ relation is found, in agreement with the few other measurements (e.g., Milky Way, M32, NGC 4258, NGC 7457) in this σ range.

For pseudobulge galaxies (see Fig. 7.2b) there is no indication for an offset as suggested by (Hu, 2008). NGC 3368 and NGC 3489 are both in good agreement with the $M_\bullet$-σ relation, aside from the variety of σ measurements in the literature for NGC 3368 (see Section 6.6 for a critical discussion). The same is true for barred galaxies. Both NGC 3368 and NGC 3489 have a bar, and therefore strengthen the evidence from Fig. 1.3d that barred galaxies are not offset from the $M_\bullet$-σ relation, in contrast to the findings of Graham (2008a); Graham & Li (2009).

Fornax A is, after Cen A (e.g., Cappellari et al. 2009), only the second merger remnant with a measured black hole mass. Both follow the $M_\bullet$-σ relation quite well (Fig. 7.2c). This supports the results of merger simulations (e.g. Johansson et al. 2009), where a galaxy after a gas-rich, mixed or dry merger event follows the $M_\bullet$-σ relation.

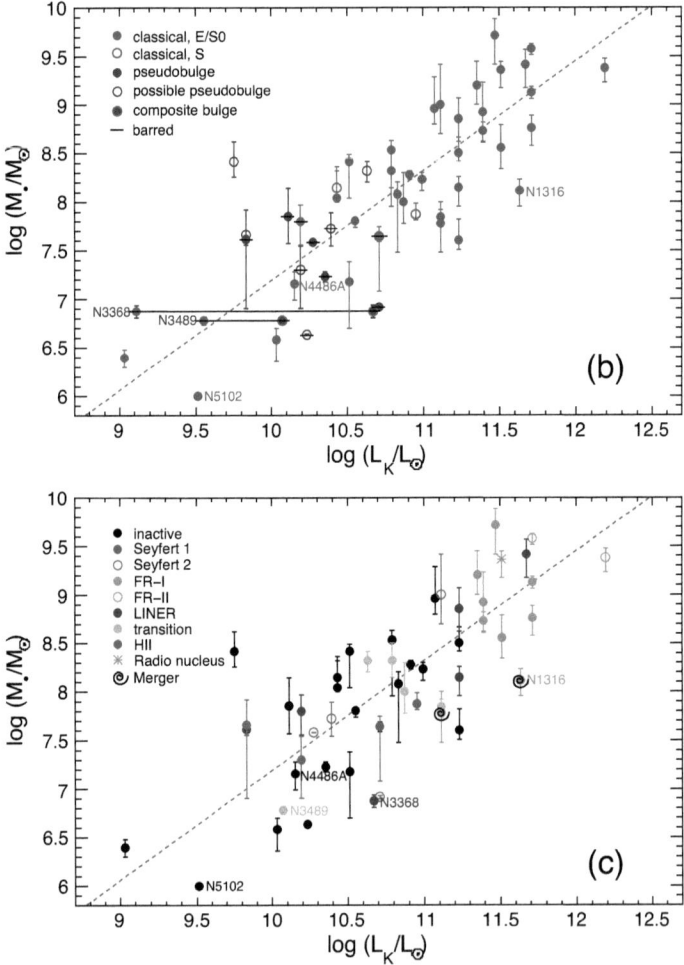

Figure 7.3: Continued. Updated $M_\bullet$-L_K relation for relevant subsamples. The galaxies analysed in Chapters 4-6 are marked by their names. (b) Classical bulges (red) and pseudobulges (blue). For the galaxies with a composite bulge, NGC 3368 and NGC 3489, two data points are plotted: L_K for the entire bulge (blue with red border) and L_K for the classical component only (red). (c) Inactive galaxies (black) and different types of AGN (coloured).

CHAPTER 7. CONCLUDING REMARKS

Implications for the $M_\bullet$-L_{bulge} relation

Fig. 7.3a shows the location of the galaxies analysed in this thesis with respect to the $M_\bullet$-L_K relation of Marconi & Hunt (2003). Most galaxies seem to be offset to lower black hole masses than indicated by the Marconi & Hunt relation. Only for low-mass classical bulges the offset does not seem to be significant. NGC 4486a agrees well with the relation. NGC 5102 seems to be an outlier, but the K-band magnitude used here is only a preliminary value, which is not based on a proper 2-dimensional bulge-disc decomposition, as would be indicated for this galaxy, and is therefore very likely overestimated.

For the composite pseudobulges the situation is more complicated. When using the K-band magnitude of the photometric bulge region (i.e., pseudobulge and classical bulge together) of NGC 3368 and NGC 3489, both galaxies fall far below the $M_\bullet$-L_K relation of Marconi & Hunt (2003) (factor of ~ 12 in the case of NGC 3368 and ~ 3 in the case of NGC 3489). The classical bulge component alone, however, seems to be a better indicator of $M_\bullet$ (Fig. 7.3b). As the stellar populations in the pseudobulges of NGC 3368 and NGC 3489 are quite young ($\lesssim 3$ Gyr), the ageing of the system has to be taken into account. The average age of the populations in these two galaxies is between 1 and 3 Gyr. They become fainter when they age passively. According to the models of Maraston (1998) and Maraston (2005), the magnitude difference in the K band between 1 – 3 Gyr and 10 Gyr is 0.9 – 2.3, which would move the pseudobulges more in the direction of the $M_\bullet$-L_K relation. Due to the large uncertainties in the age estimate, the exact amplitude of the effect is unknown.

Fornax A, as well as the other merger remnant Cen A, also both fall clearly below the $M_\bullet$-L_K relation (Fig. 7.3c). Due to the young age of the stellar populations in Fornax A ($\sim 2-3$ Gyr), the effect of ageing also has to be taken into account. The magnitude difference between 2 and 10 Gyr is only 1.4, which would move Fornax A a bit towards the $M_\bullet$-L_K relation, but still it would be a factor ~ 2 below. The stellar population of Cen A is older (~ 5 Gyr, Peng et al. 2004), thus the effect of ageing would be much less significant than for Fornax A. Simulations of Johansson et al. (2009) imply an overall good agreement of different types of merger remnants with the $M_\bullet$-L_{bulge} relation with possibly a slightly steeper slope, which might be sensitive to the initial conditions. Thus age does not seem to explain the amplitude of the offset alone, and evolutionary aspects might play a role as well. More SMBH mass measurements in merger remnants would certainly be helpful in order to understand if the offset of Fornax A and Cen A from the $M_\bullet$-L_{bulge} relationship is purely statistical, or whether it is characteristic for merger remnants.

The use of scaling relations as $M_\bullet$ estimators

In summary, the $M_\bullet$-σ relation seems to be a very good estimator of $M_\bullet$ for all galaxy types, and, compared to the $M_\bullet$-$L_{\rm bulge}$ relation, it is the more fundamental one and thus better suited to constrain theoretical models of black hole and galaxy formation and evolution. Part of the scatter of the $M_\bullet$-σ relationship might simply be due to uncertainties in σ measurements, as for many galaxies discrepant σ measurements can be found in the literature. The $M_\bullet$-$L_{\rm bulge}$ relationship is certainly useful as a $M_\bullet$ estimator for elliptical galaxies and spirals with classical bulges, but cannot be non-restrictively used for all galaxies. Apart from the generally larger scatter, especially galaxies with a pseudobulge, young stellar populations and signs of recent merger events might be displaced from the $M_\bullet$-$L_{\rm bulge}$ relation. Doing a proper pseudobulge-classical bulge decomposition and the correction for ageing might help, but this is connected with a lot of work and the involvement of more uncertainties, as the age is usually not known very accurately and a decomposition is only possible for nearby galaxies with high-resolution imaging. For galaxies at higher redshifts, where the fraction of merger galaxies is higher and the stellar populations are younger, the $M_\bullet$-$L_{\rm bulge}$ relation might give a biased result. Thus in particular when studying the black hole population in large galaxy samples that include AGN or high-redshift objects, it is advisable to use the velocity dispersion as a $M_\bullet$ indicator instead of the bulge magnitude.

Status and outlook

Presently 32 galaxies were observed with SINFONI for this project (see Table 4.1 in Chapter 2):

- eleven high-σ galaxies ($\sigma > 300$ km s^{-1}), two of them have a core
- four core ellipticals, two of them with $\sigma > 300$ km s^{-1}
- nine low-σ galaxies ($\sigma < 130$ km s^{-1}), six of them have a pseudobulge
- eleven pseudobulges, six of them with $\sigma < 130$ km s^{-1}
- three merger remnants
- two galaxies with a classical bulge

Bars are present in $\sim 40\%$ of the observed galaxies. The data reduction is ongoing, with about one third of the data readily reduced. Apart from the four galaxies discussed in this work, the stellar dynamical modelling of NGC 5102 (Thomas et al., in preparation) and NGC 1332 (Rusli et al., in preparation) is almost finished. Based on the experience collected with the reduction, analysis and dynamical modelling of the galaxies in this work, it is expected that the remaining galaxies can be processed on a much shorter timescale.

Due to the large number of pseudobulge galaxies in the sample, presumably all open questions regarding the location of pseudobulges in the $M_\bullet$-σ and $M_\bullet$-L_{bulge} relations will be answered, which is particularly important for understanding the influence of different growing mechanisms (merger, secular evolution) on the co-evolution of black holes and bulges.

As approximately half of the observed galaxies are barred, the question whether barred galaxies are offset from the $M_\bullet$-σ relation (Graham, 2008a; Graham & Li, 2009) or not (cf. Chapter 1) can probably be answered as well. As barred galaxies are not axisymmetric, it is planned to model at least some of the barred galaxies also with the triaxial code NMAGIC (de Lorenzi et al., 2007). This code, however, is not yet ready to use for the given problem (black holes in barred galaxies). Presently it is tested on the non-barred, axisymmetric galaxy NGC 4486a and the resulting models are compared with the models from the axisymmetric Schwarzschild code, so in this way the new code can be improved (de Lorenzi & Thomas, in preparation). Modelling barred galaxies with both an axisymmetric and a triaxial code will help to identify possible systematic errors and biases introduced by the simplifications made by the axisymmetric code.

The large number of high-σ galaxies will improve the statistics at the high-mass end and hopefully help to identify a breakdown, change of slope or increasing scatter of the $M_\bullet$-σ relation if present. The observed core galaxies will help to understand the imprint of core scouring on the orbital structure and strengthen the relation between the mass deficit and $M_\bullet$ found by Kormendy & Bender (2009).

Nine galaxies have been observed at the low-mass end between $\sigma = 60$ km s^{-1} and $\sigma = 130$ km s^{-1}, thereof six pseudobulges. These galaxies will improve the statistics at the low-mass end and help to tighten the slope of the $M_\bullet$-σ relation, which at the moment varies roughly between 4 and 5, depending on the used galaxy sample. A change of slope or breakdown of the $M_\bullet$-σ relation does not seem very likely in this σ range given the present data, but might happen in the intermediate-mass black hole range ($< 10^6$ $M_\bullet$), where possibly seed black holes in the earliest evolutionary stages as well as bulgeless galaxies without a central black hole are located. Unfortunately galaxies with a very low velocity dispersion and a low-mass black hole cannot be observed with SINFONI, as both the spectral and the spatial resolution is not high enough. This will be a task for the next generation of telescopes and instruments.

About one third of the observed galaxies shows some sort of weak nuclear activity, and three galaxies (including Fornax A) show a recent merger history. The observed merger remnants will hopefully give more insights on the relations between merger type, bulge growth and black hole growth. However, due to the small number and weakness of the AGN no new conclusions regarding the connection between AGN activity, bulge growth and black hole growth are expected from the observed galaxy sample. From the so far dynamically studied AGN (Figs. 7.2b and 7.3b) there do not seem to be obvious differences to the $M_\bullet$-bulge relations of inactive galaxies, but e.g. Greene & Ho (2006) suggest some deviations based on a sample of local SDSS AGN. More stellar dynamical studies of black holes in AGN would be important in many respects, e.g. to confirm the validity of reverberation mapping measurements and thus constrain the properties of the broad line region, or to confirm the validity of the $M_\bullet$-σ and $M_\bullet$-L_{bulge} relations such that these relations can be used appropriately for studies of AGN at higher redshifts. However, strong AGN are rare within the distance range, where the sphere of influence can be resolved by SINFONI. The only few suitable objects have been studied by Davies et al. (2006), Onken et al. (2007) and Hicks & Malkan (2008). Despite being a very interesting and important topic, there is presently not much to add given the technical limitations. Thus, as in the case of the extreme low-mass end, this will be a task for the next generation of telescopes and instruments.

Bibliography

Abuter R., Schreiber J., Eisenhauer F., Ott T., Horrobin M., Gillessen S., 2006, *SINFONI data reduction software*, New Astronomy Review, 50, 398 34, 51, 91, 108, 163

Ádámkovics M., de Pater I., Hartung M., Eisenhauer F., Genzel R., Griffith C. A., 2006, *Titan's bright spots: Multiband spectroscopic measurement of surface diversity and hazes*, Journal of Geophysical Research (Planets), 111, 7 44

Aller M. C., Richstone D., 2002, *The Cosmic Density of Massive Black Holes from Galaxy Velocity Dispersions*, AJ, 124, 3035 5

Aller M. C., Richstone D. O., 2007, *Host Galaxy Bulge Predictors of Supermassive Black Hole Mass*, ApJ, 665, 120 11, 24

Arnaboldi M., Freeman K. C., Gerhard O., Matthias M., Kudritzki R. P., Méndez R. H., Capaccioli M., Ford H., 1998, *The Stellar Dynamics and Mass of NGC 1316 Using the Radial Velocities of Planetary Nebulae*, ApJ, 507, 759 111

Athanassoula E., 2005, *On the nature of bulges in general and of box/peanut bulges in particular: input from N-body simulations*, MNRAS, 358, 1477 17, 146

Atkinson J. W., et al., 2005, *Supermassive black hole mass measurements for NGC 1300 and NGC 2748 based on Hubble Space Telescope emission-line gas kinematics*, MNRAS, 359, 504 15

Baes M., et al., 2003, *Radiative transfer in disc galaxies - III. The observed kinematics of dusty disc galaxies*, MNRAS, 343, 1081 185

Barberà C., Athanassoula E., García-Gómez C., 2004, *Deprojecting spiral galaxies using Fourier analysis. Application to the Frei sample*, A&A, 415, 849 156

Barth A., Ho L., Rutledge R., Sargent W., 2004, *POX 52: A Dwarf Seyfert 1 Galaxy with an Intermediate-Mass Black Hole*, ApJ, 607, 90 15

Bibliography

Barth A. J., Ho L. C., Sargent W. L. W., 2002, *A Study of the Direct Fitting Method for Measurement of Galaxy Velocity Dispersions*, AJ, 124, 2607 197

Barth A. J., Sarzi M., Rix H.-W., Ho L. C., Filippenko A. V., Sargent W. L. W., 2001, *Evidence for a Supermassive Black Hole in the S0 Galaxy NGC 3245*, ApJ, 555, 685 15, 149

Barth A. J., Strigari L. E., Bentz M. C., Greene J. E., Ho L. C., 2009, *Dynamical Constraints on the Masses of the Nuclear Star Cluster and Black Hole in the Late-Type Spiral Galaxy NGC 3621*, ApJ, 690, 1031 15

Batcheldor D., et al., 2005, *Integral Field Spectroscopy of 23 Spiral Bulges*, ApJS, 160, 76 15, 24

Bedregal A. G., Aragón-Salamanca A., Merrifield M. R., Milvang-Jensen B., 2006, *S0 galaxies in Fornax: data and kinematics*, MNRAS, 371, 1912 111, 131, 142

Beifiori A., Sarzi M., Corsini E. M., Dalla Bonta E., Pizzella A., Coccato L., Bertola F., 2009, *Upper Limits on the Masses of 105 Supermassive Black Holes from Hubble Space Telescope/Space Telescope Imaging Spectrograph Archival Data*, ApJ, 692, 856 184

Beletsky Y., Moiseev A., Alves J., Kniazev A., 2009, *The inner structure of the merger galaxy NGC 1316 (Fornax A): a normal boxy elliptical with a kinematically decoupled core*, submitted to A&A 111, 114

Bender R., 1990, *Unraveling the Kinematics of Early-Type Galaxies*, A&A, 229, 441 61, 70

Bender R., et al., 2005, *HST STIS Spectroscopy of the Triple Nucleus of M31: Two Nested Disks in Keplerian Rotation around a Supermassive Black Hole*, ApJ, 631, 280 15, 169

Bender R., Kormendy J., 2003, *Supermassive Black Holes in Galaxy Centers*, in *Astronomy, Cosmology and Fundamental Physics*, edited by Shaver P., Dilella L., Giménez A., Springer Verlag, p. 262 90

Bender R., Moellenhoff C., 1987, *Morphological analysis of massive early-type galaxies in the Virgo cluster*, A&A, 177, 71 113

Bender R., Saglia R., Gerhard O., 1994, *Line-of-Sight Velocity Distributions of Elliptical Galaxies*, MNRAS, 269, 785 111, 168

Bernardi M., et al., 2006, *A Search for the Most Massive Galaxies: Double Trouble?*, AJ, 131, 2018 25, 26

Bibliography

Best P. N., Kauffmann G., Heckman T. M., Brinchmann J., Charlot S., Ivezić Ž., White S. D. M., 2005, *The host galaxies of radio-loud active galactic nuclei: mass dependences, gas cooling and active galactic nuclei feedback*, MNRAS, 362, 25 30

Beuing J., Bender R., Mendes de Oliveira C., Thomas D., Maraston C., 2002, *Line-strength indices and velocity dispersions for 148 early-type galaxies in different environments*, A&A, 395, 431 123

Binney J., 1978, *On the rotation of elliptical galaxies*, MNRAS, 183, 501 17

Binney J., Merrifield M., 1998, *Galactic astronomy*. Princeton Series in Astrophysics, Princeton University Press 114

Böker T., Laine S., van der Marel R. P., Sarzi M., Rix H.-W., Ho L. C., Shields J. C., 2002, *A Hubble Space Telescope Census of Nuclear Star Clusters in Late-Type Spiral Galaxies. I. Observations and Image Analysis*, AJ, 123, 1389 39

Bonaccini D., et al., 2002, *ESO VLT laser guide star facility*, in *Adaptive Optics Systems and Technology II*, edited by Tyson R. K., Bonaccini D., Roggemann M. C., Vol. 4494 of Proc. SPIE, pp 276–289 38, 163

Bonnet H., et al., 2003, *Implementation of MACAO for SINFONI at the VLT, in NGS and LGS modes*, in *Adaptive Optical System Technologies II*, edited by Wizinowich P., Bonaccini D., Vol. 4839 of Proc. SPIE, pp 329–343 34, 91

Bonnet H., et al., 2004, *SINFONI – Integral field spectroscopy at 50 milli-arcsecond resolution with the ESO-VLT*, ESO Messenger, 117, 17 5, 34, 90, 91, 107, 157

Bournaud F., Combes F., Jog C. J., Puerari I., 2005, *Lopsided spiral galaxies: evidence for gas accretion*, A&A, 438, 507 169

Bower G. A., et al., 1998, *Kinematics of the Nuclear Ionized Gas in the Radio Galaxy M84 (NGC 4374)*, ApJ, 492, L111 15

Bower G. A., et al., 2001, *Evidence of a Supermassive Black Hole in the Galaxy NGC 1023 from the Nuclear Stellar Dynamics*, ApJ, 550, 75 15

Bromm V., Larson R. B., 2004, *The First Stars*, A&AR, 42, 79 3

Bureau M., Athanassoula E., 2005, *Bar Diagnostics in Edge-On Spiral Galaxies. III. N-Body Simulations of Disks*, ApJ, 626, 159 185

Bibliography

Burkert A., Silk J., 2001, *Star Formation-Regulated Growth of Black Holes in Protogalactic Spheroids*, ApJ, 554, L151 6, 90

Caon N., Capaccioli M., D'Onofrio M., 1994, *'Global mapping' B-band photometry of a complete sample of Fornax and Virgo early-type galaxies*, A&AS, 106, 199 107

Caon N., Macchetto D., Pastoriza M., 2000, *A Survey of the Interstellar Medium in Early-Type Galaxies. III. Stellar and Gas Kinematics*, ApJS, 127, 39 174

Capetti A., Marconi A., Macchetto D., Axon D., 2005, *The supermassive black hole in the Seyfert 2 galaxy NGC 5252*, A&A, 431, 465 15

Cappellari M., Copin Y., 2003, *Adaptive spatial binning of integral-field spectroscopic data using Voronoi tessellations*, MNRAS, 342, 345 83, 170

Cappellari M., Emsellem E., 2004, *Parametric Recovery of Line-of-Sight Velocity Distributions from Absorption-Line Spectra of Galaxies via Penalized Likelihood*, PASP, 116, 138 64

Cappellari M., et al., 2006, *The SAURON project – IV. The mass-to-light ratio, the virial mass estimator and the Fundamental Plane of elliptical and lenticular galaxies*, MNRAS, 366, 1126 133

Cappellari M., et al., 2008, *Supermassive black holes from OASIS and SAURON integral-field kinematics*, in *Formation and Evolution of Galaxy Bulges*, edited by Bureau M., Athanassoula E., Barbuy B., Proc. IAU Symposium 245, pp 215–218 26

Cappellari M., Neumayer N., Reunanen J., van der Werf P. P., de Zeeuw P. T., Rix H.-W., 2009, *The mass of the black hole in Centaurus A from SINFONI AO-assisted integral-field observations of stellar kinematics*, MNRAS, p. 175 11, 15, 31, 106, 143, 205

Cappellari M., Verolme E. K., van der Marel R. P., Kleijn G. A. V., Illingworth G. D., Franx M., Carollo C. M., de Zeeuw P. T., 2002, *The Counterrotating Core and the Black Hole Mass of IC 1459*, ApJ, 578, 787 15

Carollo C., Stiavelli M., Mack J., 1998, *Spiral Galaxies with WFPC2. II. The Nuclear Properties of 40 Objects*, AJ, 116, 68 39

Carollo C. M., Franx M., Illingworth G. D., Forbes D. A., 1997, *Ellipticals with Kinematically Distinct Cores: V-I Color Images with WFPC2*, ApJ, 481, 710 22, 149

Casares J., 2007, *Observational evidence for stellar-mass black holes*, in *Black Holes from Stars to Galaxies – Across the Range of Masses*, edited by Karas V., Matt G., Proc. IAU Symposium 238, pp 3–12 2

Cesetti M., et al., 2009, *Early-type galaxies in the near-infrared: 1.5-2.4 μm spectroscopy*, A&A, 497, 41 172

Chandrasekhar S., 1931, *The Maximum Mass of Ideal White Dwarfs*, ApJ, 74, 81 2

Comerón S., Knapen J. H., Beckman J. E., 2008, *On the morphology of sigma-drop galaxies*, A&A, 485, 695 166

Copin Y., Cretton N., Emsellem E., 2004, *Axisymmetric dynamical models for SAURON and OASIS observations of NGC 3377*, A&A, 415, 889 15

Corsini E. M., Wegner G., Saglia R. P., Thomas J., Bender R., Thomas D., 2008, *Spatially Resolved Spectroscopy of Coma Cluster Early-Type Galaxies. IV. Completing the Data Set*, ApJS, 175, 462 113

Cretton N., van den Bosch F. C., 1999, *Evidence for a Massive Black Hole in the S0 Galaxy NGC 4342*, ApJ, 514, 704 15

Croton D. J., et al., 2006, *The many lives of active galactic nuclei: cooling flows, black holes and the luminosities and colours of galaxies*, MNRAS, 365, 11 30

Dale D. A., Sheth K., Helou G., Regan M. W., Hüttemeister S., 2005, *Warm and Cold Molecular Gas in Galaxies*, AJ, 129, 2197 185

Dalla Bontà E., Ferrarese L., Corsini E. M., Miralda-Escudé J., Coccato L., Sarzi M., Pizzella A., Beifiori A., 2009, *The High-Mass End of the Black Hole Mass Function: Mass Estimates in Brightest Cluster Galaxies*, ApJ, 690, 537 15

Dalle Ore C., Faber S. M., Jesus J., Stoughton R., Burstein D., 1991, *Galaxy velocity dispersions using a cross-correlation method*, ApJ, 366, 38 197

Davidge T. J., 2008, *The Stellar Content of the Post-Starburst S0 Galaxy NGC 5102*, AJ, 135, 1636 68

Davies R., 2008, *AO Assisted Spectroscopy with SINFONI: PSF, Background, and Interpolation*, in *The 2007 ESO Instrument Calibration Workshop*, edited by A. Kaufer & F. Kerber, pp 249–258 50

Davies R. I., 2007, *A method to remove residual OH emission from near-infrared spectra*, MNRAS, 375, 1099 51, 56

Davies R. I., et al., 2006, *The Star-forming Torus and Stellar Dynamical Black Hole Mass in the Seyfert 1 Nucleus of NGC 3227*, ApJ, 646, 754 10, 15, 50, 99, 108, 125, 210

Davies R. I., Mueller Sánchez F., Genzel R., Tacconi L. J., Hicks E. K. S., Friedrich S., Sternberg A., 2007, *A Close Look at Star Formation around Active Galactic Nuclei*, ApJ, 671, 1388 68, 108, 123, 168, 170, 172

Davies R. I., Tacconi L., Genzel R., Ott T., Rabien S., 2004, *Using adaptive optics to probe the dynamics and star formation in active galactic nuclei*, in *Advancements in Adaptive Optics*, edited by Bonaccini D., Ellerbroek B. L., Ragazzoni R., Vol. 5490 of Proc. SPIE, pp 473–482 50, 108

Davies R. L., 2000, *Science & Surveys with Integral Field Spectrographs (Review)*, in *Imaging the Universe in Three Dimensions*, edited by van Breugel W., Bland-Hawthorn J., Vol. 195 of Astronomical Society of the Pacific Conference Series, p. 134 130

Davies R. L., Burstein D., Dressler A., Faber S. M., Lynden-Bell D., Terlevich R. J., Wegner G., 1987, *Spectroscopy and photometry of elliptical galaxies. II - The spectroscopic parameters*, ApJS, 64, 581 23

de Francesco G., Capetti A., Marconi A., 2006, *Measuring supermassive black holes with gas kinematics: the active S0 galaxy NGC 3998*, A&A, 460, 439 15

de Francesco G., Capetti A., Marconi A., 2008, *Measuring supermassive black holes with gas kinematics. II. The LINERs IC 989, NGC 5077, and NGC 6500*, A&A, 479, 355 15

de Lorenzi F., Debattista V. P., Gerhard O., Sambhus N., 2007, *NMAGIC: a fast parallel implementation of a χ^2-made-to-measure algorithm for modelling observational data*, MNRAS, 376, 71 9, 198, 209

de Lorenzo-Cáceres A., Falcón-Barroso J., Vazdekis A., Martínez-Valpuesta I., 2008, *Stellar Kinematics in Double-Barred Galaxies: The σ-Hollows*, ApJ, 684, L83 170

Devereux N., Ford H., Tsvetanov Z., Jacoby G., 2003, *STIS Spectroscopy of the Central 10 Parsecs of M81: Evidence for a Massive Black Hole*, AJ, 125, 1226 15

Di Matteo T., Springel V., Hernquist L., 2005, *Energy input from quasars regulates the growth and activity of black holes and their host galaxies*, Nature, 433, 604 6, 106

Dong X. Y., De Robertis M. M., 2006, *Low-Luminosity Active Galaxies and Their Central Black Holes*, AJ, 131, 1236 15

Dressler A., 1979, *The dynamics and structure of the cD galaxy in Abell 2029*, ApJ, 231, 659 61

Bibliography

Dressler A., 1989, *Observational Evidence for Supermassive Black Holes*, in *Active Galactic Nuclei*, edited by Osterbrock D. E., Miller J. S., Proc. IAU Symposium 134, p. 217 4

Drory N., Fisher D. B., 2007, *A Connection between Bulge Properties and the Bimodality of Galaxies*, ApJ, 664, 640 22, 39, 152

Dumas C., 2008, *Very Large Telescope SINFONI User Manual*. European Southern Observatory 35, 36

Einstein A., 1915, *Zur allgemeinen Relativitätstheorie*, Sitzungsberichte der Königlich Preußischen Akademie der Wissenschaften (Berlin), pp 778–786 2

Eisenhauer F., et al., 2003a, *SINFONI – Integral field spectroscopy at 50 milli-arcsecond resolution with the ESO VLT*, in *Instrument Design and Performance for Optical/Infrared Ground-based Telescopes*, edited by Iye M., Moorwood A., Vol. 4841 of Proc. SPIE, pp 1548–1561 5, 34, 90, 91, 107, 157

Eisenhauer F., et al., 2003b, *The Universe in 3D: First Observations with SPIFFI, the Infrared Integral Field Spectrometer for the VLT*, ESO Messenger, 113, 17 34, 91

Emsellem E., Dejonghe H., Bacon R., 1999, *Dynamical models of NGC 3115*, MNRAS, 303, 495 15

Emsellem E., et al., 2004, *The SAURON project – III. Integral-field absorption-line kinematics of 48 elliptical and lenticular galaxies*, MNRAS, 352, 721 157, 174, 194

Erwin P., 2004, *Double-barred galaxies. I. A catalog of barred galaxies with stellar secondary bars and inner disks*, A&A, 415, 941 22, 39, 147, 150

Erwin P., 2008, *The coexistence of classical bulges and disky pseudobulges in early-type disk galaxies*, in *Formation and Evolution of Galaxy Bulges* Proc. IAU Symposium 245, pp 113–116 15, 17, 22, 146

Erwin P., Beltrán J. C. V., Graham A. W., Beckman J. E., 2003, *When Is a Bulge Not a Bulge? Inner Disks Masquerading as Bulges in NGC 2787 and NGC 3945*, ApJ, 597, 929 17, 22, 146, 150

Erwin P., Graham A. W., Caon N., 2004, *The Correlation between Supermassive Black Hole Mass and the Structure of Ellipticals and Bulges*, in *Coevolution of Black Holes and Galaxies*, edited by Ho L. C., Carnegie Observatories Astrophysics Series 16

Erwin P., Sparke L. S., 1999, *Triple Bars and Complex Central Structures in Disk Galaxies*, ApJ, 521, L37 150

Erwin P., Sparke L. S., 2003, *An Imaging Survey of Early-Type Barred Galaxies*, ApJS, 146, 299 157

Bibliography

Fabbiano G., 2004, *Ultraluminous X-ray Sources: an Observational Review*, RevMexAA, 20, 46 3

Fabbiano G., Fassnacht C., Trinchieri G., 1994, *High resolution optical and UV observations of the centers of NGC 1316 and NGC 3998 with the Hubble Space Telescope*, ApJ, 434, 67 123, 129, 130

Faber S. M., et al., 1997, *The Centers of Early-Type Galaxies with HST. IV. Central Parameter Relations*, AJ, 114, 1771 26

Faber S. M., Jackson R. E., 1976, *Velocity dispersions and mass-to-light ratios for elliptical galaxies*, ApJ, 204, 668 17

Fanaroff B. L., Riley J. M., 1974, *The morphology of extragalactic radio sources of high and low luminosity*, MNRAS, 167, 31P 30

Ferrarese L., Ford H., 2005, *Supermassive Black Holes in Galactic Nuclei: Past, Present and Future Research*, Space Science Reviews, 116, 523 11, 12, 15, 16, 18, 197, 199, 201, 203

Ferrarese L., Ford H. C., 1999, *Nuclear Disks of Gas and Dust in Early-Type Galaxies and the Hunt for Massive Black Holes: Hubble Space Telescope Observations of NGC 6251*, ApJ, 515, 583 15

Ferrarese L., Ford H. C., Jaffe W., 1996, *Evidence for a Massive Black Hole in the Active Galaxy NGC 4261 from Hubble Space Telescope Images and Spectra*, ApJ, 470, 444 15

Ferrarese L., Merritt D., 2000, *A Fundamental Relation between Supermassive Black Holes and their Host Galaxies*, ApJ, 539, L9 4, 11, 12, 90, 106, 146

Filippenko A., Ho L., 2003, *A Low-Mass Central Black Hole in the Bulgeless Seyfert 1 Galaxy NGC 4395*, ApJ, 588, L13 15, 25

Fisher D. B., 2006, *Central Star Formation and PAH Profiles in Pseudobulges and Classical Bulges*, ApJ, 642, L17 22

Fisher D. B., Drory N., 2008, *The Structure of Classical Bulges and Pseudobulges: the Link Between Pseudobulges and Sérsic Index*, AJ, 136, 773 15, 22, 39

Fisher D. B., Drory N., Fabricius M. H., 2009, *Bulges of Nearby Galaxies with Spitzer: The Growth of Pseudobulges in Disk Galaxies and its Connection to Outer Disks*, ApJ, 697, 630 22, 39

Förster Schreiber N. M., 2000, *Moderate-Resolution Near-Infrared Spectroscopy of Cool Stars: A New K-Band Library*, AJ, 120, 2089 66

Franx M., Illingworth G., Heckman T., 1989, *Major and minor axis kinematics of 22 ellipticals*, ApJ, 344, 613 61

Frei Z., Guhathakurta P., Gunn J. E., Tyson J. A., 1996, *A Catalog of Digital Images of 113 Nearby Galaxies*, AJ, 111, 174 156

Frogel J. A., Stephens A., Ramírez S., DePoy D. L., 2001, *An Accurate, Easy-To-Use Abundance Scale for Globular Clusters Based on 2.2 Micron Spectra of Giant Stars*, AJ, 122, 1896 66

Gadotti D. A., Kauffmann G., 2009, *The growth of supermassive black holes in pseudo-bulges, classical bulges and elliptical galaxies*, MNRAS, 399, 621 22, 147, 198

Gaffney N. I., Lester D. F., Doppmann G., 1995, *Measuring stellar kinematics in galaxies with the near-infrared (2 − 0) ^{12}CO absorption bandhead*, PASP, 107, 68 65

Gebhardt K., et al., 2000a, *Axisymmetric, three-integral models of galaxies: a massive black hole in NGC 3379*, AJ, 119, 1157 15, 63, 68, 70, 93, 94, 99, 100, 102, 114, 116, 130, 136, 140, 166, 178

Gebhardt K., et al., 2000b, *A Relationship between Nuclear Black Hole Mass and Galaxy Velocity Dispersion*, ApJ, 539, L13 4, 11, 12, 90, 106, 146

Gebhardt K., et al., 2001, *M33: A Galaxy with No Supermassive Black Hole*, AJ, 122, 2469 15, 25, 90

Gebhardt K., et al., 2003, *Axisymmetric Dynamical Models of the Central Regions of Galaxies*, ApJ, 583, 92 12, 15, 26, 90, 94, 99, 103, 130, 178

Gebhardt K., et al., 2007, *The Black Hole Mass and Extreme Orbital Structure in NGC 1399*, ApJ, 671, 1321 15, 121, 168

Gebhardt K., Rich R. M., Ho L. C., 2005, *An Intermediate-Mass Black Hole in the Globular Cluster G1: Improved Significance from New Keck and Hubble Space Telescope Observations*, ApJ, 634, 1093 3, 15, 25

Gebhardt K., Thomas J., 2009, *The Black Hole Mass, Stellar Mass-to-Light Ratio, and Dark Halo in M87*, ApJ, 700, 1690 15

Geldzahler B. J., Fomalont E. B., 1984, *Radio observations of the jet in Fornax A*, AJ, 89, 1650 106, 123

Gerhard O., 1993, *Line-of-Sight Velocity Profiles in Spherical Galaxies: Breaking the Degeneracy between Anisotropy and Mass*, MNRAS, 265, 213 60, 94, 111, 166

Ghez A. M., Salim S., Hornstein S. D., Tanner A., Lu J. R., Morris M., Becklin E. E., Duchêne G., 2005, *Stellar Orbits around the Galactic Center Black Hole*, ApJ, 620, 744 7

Gillessen S., Eisenhauer F., Trippe S., Alexander T., Genzel R., Martins F., Ott T., 2009, *Monitoring Stellar Orbits Around the Massive Black Hole in the Galactic Center*, ApJ, 692, 1075 7, 15, 59, 90, 103

Gössl C. A., Riffeser A., 2002, *Image reduction pipeline for the detection of variable sources in highly crowded fields*, A&A, 381, 1095 109

Goudfrooij P., Alonso M. V., Maraston C., Minniti D., 2001, *The star cluster system of the 3-Gyr-old merger remnant NGC 1316: clues from optical and near-infrared photometry*, MNRAS, 328, 237 106, 121

Goudfrooij P., Mack J., Kissler-Patig M., Meylan G., Minniti D., 2001, *Kinematics, ages and metallicities of star clusters in NGC 1316: a 3-Gyr-old merger remnant*, MNRAS, 322, 643 126, 136

Graham A., Erwin P., Caon N., Trujillo I., 2001, *A Correlation between Galaxy Light Concentration and Supermassive Black Hole Mass*, ApJ, 563, L11 11

Graham A. W., 2007, *The black hole mass - spheroid luminosity relation*, MNRAS, 379, 711 16

Graham A. W., 2008a, *Fundamental Planes and the Barless M_{BH}-σ Relation for Supermassive Black Holes*, ApJ, 680, 143 23, 24, 205, 209

Graham A. W., 2008b, *Populating the Galaxy Velocity Dispersion: Supermassive Black Hole Mass Diagram, A Catalogue of (M_{BH}, σ) Values*, PASA, 25, 167 11, 12

Graham A. W., Driver S. P., 2007, *A Log-Quadratic Relation for Predicting Supermassive Black Hole Masses from the Host Bulge Sérsic Index*, ApJ, 655, 77 11

Graham A. W., Li I.-h., 2009, *The M_{BH}-σ Diagram and the Offset Nature of Barred Active Galaxies*, ApJ, 698, 812 205, 209

Greene J. E., Ho L. C., 2006, *The M_{BH}-σ_* Relation in Local Active Galaxies*, ApJ, 641, L21 30, 210

Greene J. E., Ho L. C., Barth A. J., 2008, *Black Holes in Pseudobulges and Spheroidals: A Change in the Black Hole-Bulge Scaling Relations at Low Mass*, ApJ, 688, 159 22, 147, 198

Greenhill L. J., et al., 2003, *A Warped Accretion Disk and Wide-Angle Outflow in the Inner Parsec of the Circinus Galaxy*, ApJ, 590, 162 15

Gu Q., Melnick J., Fernandes R. C., Kunth D., Terlevich E., Terlevich R., 2006, *Emission-line properties of Seyfert 2 nuclei*, MNRAS, 366, 480 30

Gültekin K., et al., 2009a, *A Quintet of Black Hole Mass Determinations*, ApJ, 695, 1577 15

Gültekin K., et al., 2009b, *The M-σ and M-L Relations in Galactic Bulges, and Determinations of Their Intrinsic Scatter*, ApJ, 698, 198 11, 12, 15, 16, 22, 24

Gürkan M. A., Freitag M., Rasio F. A., 2004, *Formation of Massive Black Holes in Dense Star Clusters. I. Mass Segregation and Core Collapse*, ApJ, 604, 632 3

Haan S., Schinnerer E., Emsellem E., García-Burillo S., Combes F., Mundell C. G., Rix H.-W., 2009, *Dynamical Evolution of AGN Host Galaxies – Gas In/Out-Flow Rates in Seven NUGA Galaxies*, ApJ, 692, 1623 169

Haan S., Schinnerer E., Mundell C. G., García-Burillo S., Combes F., 2008, *Atomic Hydrogen Properties of Active Galactic Nuclei Host Galaxies: HI in 16 Nuclei of Galaxies (NUGA) Sources*, AJ, 135, 232 169

Haehnelt M. G., Kauffmann G., 2000, *The correlation between black hole mass and bulge velocity dispersion in hierarchical galaxy formation models*, MNRAS, 318, L35 90

Haehnelt M. G., Rees M. J., 1993, *The formation of nuclei in newly formed galaxies and the evolution of the quasar population*, MNRAS, 263, 168 3

Häring N., Rix H.-W., 2004, *On the Black Hole Mass-Bulge Mass Relation*, ApJ, 604, L89 11, 16, 146, 147

Häring-Neumayer N., Cappellari M., Rix H.-W., Hartung M., Prieto M. A., Meisenheimer K., Lenzen R., 2006, *VLT Diffraction-limited Imaging and Spectroscopy in the NIR: Weighing the Black Hole in Centaurus A with NACO*, ApJ, 643, 226 11, 31, 106

Heckman T. M., 2008, *The Co-Evolution of Galaxies and Black Holes: Current Status and Future Prospects*, preprint (arXiv:0809.1101 [astro-ph]) 30

Heckman T. M., Kauffmann G., Brinchmann J., Charlot S., Tremonti C., White S. D. M., 2004, *Present-Day Growth of Black Holes and Bulges: The Sloan Digital Sky Survey Perspective*, ApJ, 613, 109 30

Helfer T. T., Thornley M. D., Regan M. W., Wong T., Sheth K., Vogel S. N., Blitz L., Bock D. C.-J., 2003, *The BIMA Survey of Nearby Galaxies (BIMA SONG). II. The CO Data*, ApJS, 145, 259 185, 186, 187

Bibliography

Héraudeau P., Simien F., 1998, *Stellar kinematical data for the central region of spiral galaxies. I.*, A&AS, 133, 317 15

Héraudeau P., Simien F., Maubon G., Prugniel P., 1999, *Stellar kinematic data for the central region of spiral galaxies. II.*, A&AS, 136, 509 150, 152, 168, 169, 197

Herrnstein J. R., et al., 1999, *A geometric distance to the galaxy NGC 4258 from orbital motions in a nuclear gas disk*, Nature, 400, 539 15

Hicks E. K. S., Malkan M. A., 2008, *Circumnuclear Gas in Seyfert 1 Galaxies: Morphology, Kinematics, and Direct Measurement of Black Hole Masses*, ApJS, 174, 31 10, 15, 210

Ho L., Filippenko A., Sargent W., 1997, *A Search for "Dwarf" Seyfert Nuclei. III. Spectroscopic Parameters and Properties of the Host Galaxies*, ApJS, 112, 315 26, 147, 148

Holley-Bockelmann K., Richstone D. O., 2000, *The Role of a Massive Central Singularity in Galactic Mergers on the Survival of the Core Fundamental Plane*, ApJ, 531, 232 121

Hopkins P. F., Hernquist L., Cox T. J., Kereš D., 2008, *A Cosmological Framework for the Co-Evolution of Quasars, Supermassive Black Holes, and Elliptical Galaxies. I. Galaxy Mergers and Quasar Activity*, ApJS, 175, 356 106

Horellou C., Black J. H., van Gorkom J. H., Combes F., van der Hulst J. M., Charmandaris V., 2001, *Atomic and molecular gas in the merger galaxy NGC 1316 (Fornax A) and its environment*, A&A, 376, 837 126

Houghton R. C. W., Magorrian J., Sarzi M., Thatte N., Davies R. L., Krajnović D., 2006, *The central kinematics of NGC 1399 measured with 14 pc resolution*, MNRAS, 367, 2 15, 121

Hu J., 2008, *The black hole mass-stellar velocity dispersion correlation: bulges versus pseudo-bulges*, MNRAS, 386, 2242 15, 22, 23, 147, 205

Humphrey P. J., Buote D. A., Brighenti F., Gebhardt K., Mathews W. G., 2008, *Weighing the Quiescent Central Black Hole in an Elliptical Galaxy with X-Ray-Emitting Gas*, ApJ, 683, 161 10, 15

Iyomoto N., Makishima K., Tashiro M., Inoue S., Kaneda H., Matsumoto Y., Mizuno T., 1998, *The Declined Activity in the Nucleus of NGC 1316*, ApJ, 503, L31 106, 123

Jarrett T. H., Chester T., Cutri R., Schneider S. E., Huchra J. P., 2003, *The 2MASS Large Galaxy Atlas*, AJ, 125, 525 156

Bibliography

Jog C. J., Combes F., 2009, *Lopsided spiral galaxies*, Physics Reports, 471, 75 169

Joguet B., Kunth D., Melnick J., Terlevich R., Terlevich E., 2001, *Stellar populations in Seyfert 2 galaxies. I. Atlas of near-UV spectra*, A&A, 380, 19 30

Johansson P. H., Naab T., Burkert A., 2009, *Equal- and Unequal-Mass Mergers of Disk and Elliptical Galaxies with Black Holes*, ApJ, 690, 802 106, 205, 207

Jorgensen I., Franx M., Kjaergaard P., 1995, *Spectroscopy for E and S0 galaxies in nine clusters*, MNRAS, 276, 1341 12

Joseph C. L., et al., 2001, *The Nuclear Dynamics of M32. I. Data and Stellar Kinematics*, ApJ, 550, 668 70, 72, 74, 94, 115

Karachentsev I. D., et al., 2007, *The Hubble Flow around the Centaurus A/M83 Galaxy Complex*, AJ, 133, 504 15

Kerr R. P., 1963, *Gravitational Field of a Spinning Mass as an Example of Algebraically Special Metrics*, Phys. Rev. Lett., 11, 237 2

Kim D.-W., Fabbiano G., 2003, *Chandra X-Ray Observations of NGC 1316 (Fornax A)*, ApJ, 586, 826 123, 125

King A., 2003, *Black Holes, Galaxy Formation, and the M_{BH}-σ Relation*, ApJ, 596, L27 6

King A. R., Pounds K. A., 2003, *Black hole winds*, MNRAS, 345, 657 6

Knapen J. H., de Jong R. S., Stedman S., Bramich D. M., 2003, *Structure and star formation in disc galaxies - I. Sample selection and near-infrared imaging*, MNRAS, 344, 527 149, 151, 153, 154, 177

Kormendy J., 1982, *Rotation of the bulge components of barred galaxies*, ApJ, 257, 75 17, 146, 150

Kormendy J., 1993, *Kinematics of extragalactic bulges: evidence that some bulges are really disks*, in *Galactic Bulges*, edited by Dejonghe H., Habing H. J., Vol. 153 of IAU Symposium, p. 209 11, 17, 146

Kormendy J., 2001, *Supermassive Black Holes in Galactic Nuclei*, RevMexAA, 10, 69 11, 16, 147

Kormendy J., Bender R., 2009, *Correlations between Supermassive Black Holes, Velocity Dispersions, and Mass Deficits in Elliptical Galaxies with Cores*, ApJ, 691, L142 26, 209

Kormendy J., et al., 1996, *Hubble Space Telescope Spectroscopic Evidence for a $2 \times 10^9 M_{sun}$ Black Hole in NGC 3115*, ApJ, 459, L57 15

Kormendy J., Fisher D. B., Cornell M. E., Bender R., 2009, *Structure and Formation of Elliptical and Spheroidal Galaxies*, ApJS, 182, 216 26

Kormendy J., Gebhardt K., 2001, *Supermassive Black Holes in Galactic Nuclei (Plenary Talk)*, in *AIP Conf. Proc. 586: 20th Texas Symposium on relativistic astrophysics*, p. 363 4

Kormendy J., Gebhardt K., Fisher D., Drory N., Ducchio Macchetto F., Sparks W., 2005, *The Nuclear Disk in the Dwarf Elliptical Galaxy NGC 4486A*, AJ, 129, 2636 90, 102

Kormendy J., Kennicutt, Jr. R., 2004, *Secular Evolution and the Formation of Pseudobulges in Disk Galaxies*, A&AR, 42, 603 17, 22, 24, 39, 146

Kormendy J., Richstone D., 1995, *Inward Bound: The Search for Supermassive Black Holes in Galaxy Nuclei*, A&AR, 33, 581 3, 4, 11, 16, 90, 146

Krajnović D., Cappellari M., de Zeeuw P. T., Copin Y., 2006, *Kinemetry: a generalization of photometry to the higher moments of the line-of-sight velocity distribution*, MNRAS, 366, 787 170

Krajnović D., Sharp R., Thatte N., 2007, *Integral-field spectroscopy of Centaurus A nucleus*, MNRAS, 374, 385 11, 31, 106

Kuijken K., Merrifield M. R., 1993, *A New Method for Obtaining Stellar Velocity Distributions from Absorption-Line Spectra - Unresolved Gaussian Decomposition*, MNRAS, 264, 712 63

Kuntschner H., 1998, PhD thesis, University of Durham 126, 129, 133

Kuntschner H., 2000, *The stellar populations of early-type galaxies in the Fornax cluster*, MNRAS, 315, 184 126, 133, 136

Laplace P.-S., 1796, *Exposition du Système du Monde*. Imprimerie du Cercle-Social, Paris, France 2

Lauer T. R., et al., 2007, *The Centers of Early-Type Galaxies with Hubble Space Telescope. VI. Bimodal Central Surface Brightness Profiles*, ApJ, 664, 226 15, 107

Lauer T. R., Tremaine S., Richstone D., Faber S. M., 2007, *Selection Bias in Observing the Cosmological Evolution of the $M_\bullet$-σ and $M_\bullet$-L Relationships*, ApJ, 670, 249 25

Lodato G., Bertin G., 2003, *Non-Keplerian rotation in the nucleus of NGC 1068: Evidence for a massive accretion disk?*, A&A, 398, 517 15

Longhetti M., Rampazzo R., Bressan A., Chiosi C., 1998, *Star formation history of early-type galaxies in low density environments. II. Kinematics*, A&AS, 130, 267 111

Bibliography

Lynden-Bell D., 1978, *Gravity power*, Physica Scripta, 17, 185 2

Lyubenova M., Kuntschner H., Silva D. R., 2008, *Central K-band kinematics and line strength maps of NGC 1399*, A&A, 485, 425 121, 122, 168

Macchetto F., Marconi A., Axon D. J., Capetti A., Sparks W., Crane P., 1997, *The Supermassive Black Hole of M87 and the Kinematics of Its Associated Gaseous Disk*, ApJ, 489, 579 15

Mackie G., Fabbiano G., 1998, *Evolution of gas and stars in the merger galaxy NGC 1316 (Fornax A)*, AJ, 115, 514 106, 121

Madau P., Rees M. J., 2001, *Massive Black Holes as Population III Remnants*, ApJ, 551, L27 3

Madore B. F., et al., 1999, *The Hubble Space Telescope Key Project on the Extragalactic Distance Scale. XV. A Cepheid Distance to the Fornax Cluster and Its Implications*, ApJ, 515, 29 107

Magorrian J., 1999, *Kinematical signatures of hidden stellar discs*, MNRAS, 302, 530 99, 113, 177

Magorrian J., et al., 1998, *The Demography of Massive Dark Objects in Galaxy Centers*, AJ, 115, 2285 4, 11, 16

Maiolino R., Krabbe A., Thatte N., Genzel R., 1998, *Seyfert Activity and Nuclear Star Formation in the Circinus Galaxy*, ApJ, 493, 650 12, 15

Maoz D., 2007, *Low-luminosity active galactic nuclei: are they UV faint and radio loud?*, MNRAS, 377, 1696 26, 148

Maoz D., Nagar N. M., Falcke H., Wilson A. S., 2005, *The Murmur of the Sleeping Black Hole: Detection of Nuclear Ultraviolet Variability in LINER Galaxies*, ApJ, 625, 699 148

Maraston C., 1998, *Evolutionary synthesis of stellar populations: a modular tool*, MNRAS, 300, 872 99, 102, 131, 133, 185, 194, 198, 207

Maraston C., 2005, *Evolutionary population synthesis: models, analysis of the ingredients and application to high-z galaxies*, MNRAS, 362, 799 99, 102, 131, 133, 185, 194, 198, 207

Marconi A., Capetti A., Axon D. J., Koekemoer A., Macchetto D., Schreier E. J., 2001, *Peering through the Dust: Evidence for a Supermassive Black Hole at the Nucleus of Centaurus A from VLT Infrared Spectroscopy*, ApJ, 549, 915 11, 31, 106

Marconi A., Hunt L. K., 2003, *The Relation between Black Hole Mass, Bulge Mass, and Near-Infrared Luminosity*, ApJ, 589, L21 11, 15, 16, 17, 27, 106, 143, 146, 184, 194, 197, 198, 199, 205, 207

Marconi A., Pastorini G., Pacini F., Axon D. J., Capetti A., Macchetto D., Koekemoer A. M., Schreier E. J., 2006, *The supermassive black hole in Centaurus A: a benchmark for gas kinematical measurements*, A&A, 448, 921 11, 31, 106

Mármol-Queraltó E., Cardiel N., Cenarro A. J., Vazdekis A., Gorgas J., Pedraz S., Peletier R. F., Sánchez-Blázquez P., 2008, *A new stellar library in the region of the CO index at 2.3 μm. New index definition and empirical fitting functions*, A&A, 489, 885 66, 129

Marulli F., Bonoli S., Branchini E., Moscardini L., Springel V., 2008, *Modelling the cosmological co-evolution of supermassive black holes and galaxies - I. BH scaling relations and the AGN luminosity function*, MNRAS, 385, 1846 6

Masters K. L., Springob C. M., Huchra J. P., 2008, *2MTF. I. The Tully-Fisher Relation in the Two Micron all Sky Survey J, H, and K Bands*, AJ, 135, 1738 156

McDermid R. M., et al., 2006, *The SAURON project – VIII. OASIS/CFHT integral-field spectroscopy of elliptical and lenticular galaxy centres*, MNRAS, 373, 906 174, 176, 194

McGill K. L., Woo J.-H., Treu T., Malkan M. A., 2008, *Comparing and Calibrating Black Hole Mass Estimators for Distant Active Galactic Nuclei*, ApJ, 673, 703 11

McLure R. J., Dunlop J. S., 2002, *On the black hole-bulge mass relation in active and inactive galaxies*, MNRAS, 331, 795 16

Mei S., et al., 2007, *The ACS Virgo Cluster Survey. XIII. SBF Distance Catalog and the Three-dimensional Structure of the Virgo Cluster*, ApJ, 655, 144 15

Meinel I. A. B., 1950, *OH Emission Bands in the Spectrum of the Night Sky*, ApJ, 111, 555 45

Merrifield M. R., Forbes D. A., Terlevich A. I., 2000, *The black hole mass-galaxy age relation*, MNRAS, 313, L29 143

Merritt D., 1997, *Recovering velocity distributions via penalized likelihood*, AJ, 114, 228 63, 70, 94, 115

Merritt D., Ferrarese L., 2001, *Black Hole Demographics from the $M_\bullet$-σ relation*, MNRAS, 320, L30 25

Merritt D., Ferrarese L., Joseph C. L., 2001, *No Supermassive Black Hole in M33?*, Science, 293, 1116 15

Merritt D., Mikkola S., Szell A., 2007, *Long-Term Evolution of Massive Black Hole Binaries. III. Binary Evolution in Collisional Nuclei*, ApJ, 671, 53 26

Michell J., 1784, *On the Means of Discovering the Distance, Magnitude, &c. of the Fixed Stars ...*, Royal Society of London Philosophical Transactions Series I, 74, 35 1

Miyoshi M., Moran J., Herrnstein J., Greenhill L., Nakai N., Diamond P., Inoue M., 1995, *Evidence for a Black-Hole from High Rotation Velocities in a Sub-Parsec Region of NGC 4258*, Nature, 373, 127 8, 15

Modigliani A., 2009, *Very Large Telescope SINFONI Pipeline User Manual*. European Southern Observatory, Issue 9.0 34, 51

Moiseev A. V., Valdés J. R., Chavushyan V. H., 2004, *Structure and kinematics of candidate double-barred galaxies*, A&A, 421, 433 156, 168, 169, 197

Mueller Sánchez F., Davies R. I., Eisenhauer F., Tacconi L. J., Genzel R., Sternberg A., 2006, *SINFONI adaptive optics integral field spectroscopy of the Circinus Galaxy*, A&A, 454, 481 12, 15, 50, 108, 171, 185, 187

Nelson C. H., Green R. F., Bower G., Gebhardt K., Weistrop D., 2004, *The Relationship Between Black Hole Mass and Velocity Dispersion in Seyfert 1 Galaxies*, ApJ, 615, 652 24

Nelson C. H., Whittle M., 1995, *Stellar and Gaseous Kinematics of Seyfert Galaxies. I. Spectroscopic Data*, ApJS, 99, 67 15

Neumayer N., Cappellari M., Reunanen J., Rix H.-W., van der Werf P. P., de Zeeuw P. T., Davies R. I., 2007, *The Central Parsecs of Centaurus A: High-excitation Gas, a Molecular Disk, and the Mass of the Black Hole*, ApJ, 671, 1329 11, 15, 31, 106, 142

Nowak N., Saglia R. P., Thomas J., Bender R., Davies R. I., Gebhardt K., 2008, *The supermassive black hole of Fornax A*, MNRAS, 391, 1629 78, 163, 178

Nowak N., Saglia R. P., Thomas J., Bender R., Pannella M., Gebhardt K., Davies R. I., 2007, *The supermassive black hole in NGC 4486a detected with SINFONI at the Very Large Telescope*, MNRAS, 379, 909 107, 116, 130, 178

Noyola E., Gebhardt K., Bergmann M., 2008, *Gemini and Hubble Space Telescope Evidence for an Intermediate-Mass Black Hole in ω Centauri*, ApJ, 676, 1008 3, 15, 25

Oliva E., Origlia L., Kotilainen J. K., Moorwood A. F. M., 1995, *Red supergiants as starburst tracers in galactic nuclei*, A&A, 301, 55 12, 15, 66

Oliva E., Origlia L., Maiolino R., Moorwood A. F. M., 1999, *Starbursts in active galaxy nuclei: observational constraints from IR stellar absorption lines*, A&A, 350, 9 125

Onken C. A., et al., 2007, *The Black Hole Mass of NGC 4151: Comparison of Reverberation Mapping and Stellar Dynamical Measurements*, ApJ, 670, 105 15, 24, 210

Onken C. A., Ferrarese L., Merritt D., Peterson B. M., Pogge R. W., Vestergaard M., Wandel A., 2004, *Supermassive Black Holes in Active Galactic Nuclei. II. Calibration of the Black Hole Mass-Velocity Dispersion Relationship for Active Galactic Nuclei*, ApJ, 615, 645 10, 30

Origlia L., Oliva E., 2000, *Is the [CO] index an age indicator for star forming galaxies?*, A&A, 357, 61 65

Paturel G., Petit C., Prugniel P., Theureau G., Rousseau J., Brouty M., Dubois P., Cambrésy L., 2003, *HYPERLEDA. I. Identification and designation of galaxies*, A&A, 412, 45 23

Peletier R. F., et al., 2007, *The SAURON project – XI. Stellar populations from absorption-line strength maps of 24 early-type spirals*, MNRAS, 379, 445 118, 166

Peng C. Y., Ho L. C., Impey C. D., Rix H.-W., 2002, *Detailed Structural Decomposition of Galaxy Images*, AJ, 124, 266 96

Peng E. W., Ford H. C., Freeman K. C., 2004, *The Globular Cluster System of NGC 5128. II. Ages, Metallicities, Kinematics, and Formation*, ApJ, 602, 705 207

Peng Z., Gu Q., Melnick J., Zhao Y., 2006, *The K-band properties of Seyfert 2 galaxies*, A&A, 453, 863 15

Pierini D., Gordon K. D., Witt A. N., Madsen G. J., 2004, *Dust Attenuation in Late-Type Galaxies. I. Effects on Bulge and Disk Components*, ApJ, 617, 1022 102

Portilla J. G., Rodríguez-Ardila A., Tejeiro J. M., 2008, *Near Infrared ($0.8 - 2.3$ μm) Coronal Lines in Active Galactic Nuclei*, RevMexAA, 32, 80 125

Pounds K. A., Reeves J. N., King A. R., Page K. L., O'Brien P. T., Turner M. J. L., 2003, *A high-velocity ionized outflow and XUV photosphere in the narrow emission line quasar PG1211+143*, MNRAS, 345, 705 6

Quinlan G. D., Hernquist L., 1997, *The dynamical evolution of massive black hole binaries - II. Self-consistent N-body integrations*, New Astronomy, 2, 533 26

Rabien S., Davies R. I., Ott T., Li J., Abuter R., Kellner S., Neumann U., 2004, *Test performance of the PARSEC laser system*, in *Advancements in Adaptive Optics*, edited by Bonaccini D., Ellerbroek B. L., Ragazzoni R., Vol. 5490 of Proc. SPIE, pp 981–988 38, 163

Rasio F. A., Freitag M., Gürkan M. A., 2004, *Formation of Massive Black Holes in Dense Star Clusters*, in *Coevolution of Black Holes and Galaxies*, edited by Ho L. C., p. 138 3

Rees M. J., 1998, *Astrophysical Evidence for Black Holes*, in *Black Holes and Relativistic Stars*, edited by Wald R. M., p. 79 24

Reunanen J., Kotilainen J. K., Prieto M. A., 2002, *Near-infrared spectroscopy of nearby Seyfert galaxies - I. First results*, MNRAS, 331, 154 170

Richards G. T., et al., 2006, *The Sloan Digital Sky Survey Quasar Survey: Quasar Luminosity Function from Data Release 3*, AJ, 131, 2766 3

Richstone D., et al., 1998, *Supermassive Black Holes and the Evolution of Galaxies*, Nature, 395, A14 90

Richstone D. O., Tremaine S., 1988, *Maximum-entropy models of galaxies*, ApJ, 327, 82 178

Rix H.-W., White S. D. M., 1992, *Optimal estimates of line-of-sight velocity distributions from absorption line spectra of galaxies – Nuclear discs in elliptical galaxies*, MNRAS, 254, 389 63

Rodríguez-Ardila A., Riffel R., Pastoriza M. G., 2005, *Molecular hydrogen and [FeII] in active galactic nuclei - II. Results for Seyfert 2 galaxies*, MNRAS, 364, 1041 170, 185

Saglia R. P., Maraston C., Thomas D., Bender R., Colless M., 2002, *The Puzzlingly Small CaII Triplet Absorption in Elliptical Galaxies*, ApJ, 579, L13 111, 139

Saha P., Williams T. B., 1994, *Unfolding kinematics from galaxy spectra: A Bayesian method*, AJ, 107, 1295 63

Sakamoto K., Okumura S. K., Ishizuki S., Scoville N. Z., 1999, *CO Images of the Central Regions of 20 Nearby Spiral Galaxies*, ApJS, 124, 403 156, 171, 185, 186, 187

Salpeter E. E., 1964, *Accretion of Interstellar Matter by Massive Objects.*, ApJ, 140, 796 2

Sargent W. L. W., Schechter P. L., Boksenberg A., Shortridge K., 1977, *Velocity dispersions for 13 galaxies.*, ApJ, 212, 326 61

Sarzi M., et al., 2002, *Limits on the Mass of the Central Black Hole in 16 Nearby Bulges*, ApJ, 567, 237 184, 197

Sarzi M., Rix H.-W., Shields J. C., Ho L. C., Barth A. J., Rudnick G., Filippenko A. V., Sargent W. L. W., 2005, *The Stellar Populations in the Central Parsecs of Galactic Bulges*, ApJ, 628, 169 170, 174, 176, 185, 194

Sarzi M., Rix H.-W., Shields J. C., Rudnick G., Ho L. C., McIntosh D. H., Filippenko A. V., Sargent W. L. W., 2001, *Supermassive Black Holes in Bulges*, ApJ, 550, 65 15

Schlegel D. J., Finkbeiner D. P., Davis M., 1998, *Maps of Dust Infrared Emission for Use in Estimation of Reddening and Cosmic Microwave Background Radiation Foregrounds*, ApJ, 500, 525 156

Schneider S. E., 1989, *Neutral hydrogen in the M96 group – The galaxies and the intergalactic ring*, ApJ, 343, 94 156, 169

Schödel R., et al., 2002, *A star in a 15.2-year orbit around the supermassive black hole at the centre of the Milky Way*, Nature, 419, 694 7

Schreiber J., Thatte N., Eisenhauer F., Tecza M., Abuter R., Horrobin M., 2004, *Data Reduction Software for the VLT Integral Field Spectrometer SPIFFI*, in *ASP Conf. Proc.*, edited by Ochsenbein F., Allen M., Egret D., Vol. 314, p. 380 34, 51, 91, 108, 163

Schwarzschild M., 1979, *A Numerical Model for a Triaxial Stellar System in Dynamical Equilibrium*, ApJ, 232, 236 9, 99, 130, 176

Schwarzschild K., 1916, *Über das Gravitationsfeld eines Massenpunktes nach der Einsteinschen Theorie*, Sitzungsberichte der Königlich Preussischen Akademie der Wissenschaften, 1, 189 2

Schweizer F., 1980, *An optical study of the giant radio galaxy NGC 1316 (Fornax A)*, ApJ, 237, 303 106, 121

Schweizer F., 1981, *Optical properties of the central region of NGC 1316 – A small bright core in a giant D galaxy*, ApJ, 246, 722 106, 121

Sérsic J. L., 1968, *Atlas de galaxias australes*. Córdoba, Argentina: Observatorio Astronómico 99

Shapiro K. L., Cappellari M., de Zeeuw T., McDermid R. M., Gebhardt K., van den Bosch R. C. E., Statler T. S., 2006, *The black hole in NGC 3379: a comparison of gas and stellar dynamical mass measurements with HST and integral-field data*, MNRAS, 370, 559 15

Shaya E. J., et al., 1996, *Hubble Space Telescope Planetary Camera Images of NGC 1316 (Fornax A)*, AJ, 111, 2212 122, 130

Shields G. A., Salviander S., Bonning E. W., 2006, *Evolution of the black hole - Bulge relationship in QSOs*, New Astronomy Review, 50, 809 25

Sijacki D., Springel V., di Matteo T., Hernquist L., 2007, *A unified model for AGN feedback in cosmological simulations of structure formation*, MNRAS, 380, 877 6

Sil'chenko O. K., Moiseev A. V., Afanasiev V. L., Chavushyan V. H., Valdes J. R., 2003, *The Leo I Cloud: Secular Nuclear Evolution of NGC 3379, NGC 3384, and NGC 3368?*, ApJ, 591, 185 156, 168, 169, 170, 172

Silge J., Gebhardt K., Bergmann M., Richstone D., 2005, *Gemini Near Infrared Spectrograph Observations of the Central Supermassive Black Hole in Centaurus A*, AJ, 130, 406 11, 31, 106, 143

Silge J. D., Gebhardt K., 2003, *Dust and the Infrared Kinematic Properties of Early-Type Galaxies*, AJ, 125, 2809 66, 68, 111, 115, 116, 122, 128, 166, 169

Silk J., Rees M. J., 1998, *Quasars and galaxy formation*, A&A, 331, L1 6, 90

Silva D. R., Kuntschner H., Lyubenova M., 2008, *A New Approach to the Study of Stellar Populations in Early-Type Galaxies: K-Band Spectral Indices and an Application to the Fornax Cluster*, ApJ, 674, 194 66, 80, 115, 116, 128, 129, 172

Simkin S. M., 1974, *Measurements of Velocity Dispersions and Doppler Shifts from Digitized Optical Spectra*, A&A, 31, 129 61

Siopis C., et al., 2009, *A Stellar Dynamical Measurement of the Black Hole Mass in the Maser Galaxy NGC 4258*, ApJ, 693, 946 8, 15, 24, 90, 103, 178

Skrutskie M. F., et al., 2006, *The Two Micron All Sky Survey (2MASS)*, AJ, 131, 1163 15, 41, 149

Smith R. J., Lucey J. R., Hudson M. J., Schlegel D. J., Davies R. L., 2000, *Streaming motions of galaxy clusters within 12000 km s^{-1} - I. New spectroscopic data*, MNRAS, 313, 469 197

Soltan A., 1982, *Masses of quasars*, MNRAS, 200, 115 3, 5

Springob C. M., Haynes M. P., Giovanelli R., Kent B. R., 2005, *A Digital Archive of H I 21 Centimeter Line Spectra of Optically Targeted Galaxies*, ApJS, 160, 149 156

Tadhunter C., Marconi A., Axon D., Wills K., Robinson T. G., Jackson N., 2003, *Spectroscopy of the near-nuclear regions of Cygnus A: estimating the mass of the supermassive black hole*, MNRAS, 342, 861 15

Thomas J., Jesseit R., Naab T., Saglia R. P., Burkert A., Bender R., 2007, *Axisymmetric orbit models of N-body merger remnants: a dependency of reconstructed mass on viewing angle*, MNRAS, 381, 1672 122, 185

Thomas J., Saglia R., Bender R., Thomas D., Gebhardt K., Magorrian J., Richstone D., 2004, *Mapping Stationary Axisymmetric Phase-Space Distribution Functions by Orbit Libraries*, MNRAS, 353, 391 99, 130, 178

Thomas J., Saglia R. P., Bender R., Thomas D., Gebhardt K., Magorrian J., Corsini E. M., Wegner G., 2007, *Dynamical modelling of luminous and dark matter in 17 Coma early-type galaxies*, MNRAS, 382, 657 131

Thornton C. E., Barth A. J., Ho L. C., Rutledge R. E., Greene J. E., 2008, *The Host Galaxy and Central Engine of the Dwarf Active Galactic Nucleus POX 52*, ApJ, 686, 892 15

Tonry J., Davis M., 1979, *A survey of galaxy redshifts. I. Data reduction techniques*, AJ, 84, 1511 61

Tonry J. L., Dressler A., Blakeslee J. P., Ajhar E. A., Fletcher A. B., Luppino G. A., Metzger M. R., Moore C. B., 2001, *The SBF Survey of Galaxy Distances. IV. SBF Magnitudes, Colors, and Distances*, ApJ, 546, 681 12, 15, 39, 148

Tremaine S., et al., 2002, *The Slope of the Black Hole Mass versus Velocity Dispersion Correlation*, ApJ, 574, 740 6, 11, 12, 15, 16, 18, 24, 39, 40, 42, 93, 103, 106, 107, 130, 142, 148, 168, 184, 194, 197, 199, 201, 203

Urry C. M., Padovani P., 1995, *Unified Schemes for Radio-Loud Active Galactic Nuclei*, PASP, 107, 803 9, 30

Valluri M., Ferrarese L., Merritt D., Joseph C. L., 2005, *The Low End of the Supermassive Black Hole Mass Function: Constraining the Mass of a Nuclear Black Hole in NGC 205 via Stellar Kinematics*, ApJ, 628, 137 15, 90

van den Bosch R., de Zeeuw T., Gebhardt K., Noyola E., van de Ven G., 2006, *The Dynamical Mass-to-Light Ratio Profile and Distance of the Globular Cluster M15*, ApJ, 641, 852 15 van den Bosch R. C. E., van de Ven G., Verolme E. K., Cappellari M., de Zeeuw P. T., 2008, *Triaxial orbit based galaxy models with an application to the (apparent) decoupled core galaxy NGC 4365*, MNRAS, 385, 647 9, 198

van der Marel R., Franx M., 1993, *A New Method for the Identification of Non-Gaussian Line Profiles in Elliptical Galaxies*, ApJ, 407, 525 94, 111

van der Marel R. P., 2004, *Intermediate-mass Black Holes in the Universe: A Review of Formation Theories and Observational Constraints*, in *Coevolution of Black Holes and Galaxies*, edited by Ho L. C., p. 37 3

van der Marel R. P., Cretton N., de Zeeuw P. T., Rix H.-W., 1998, *Improved Evidence for a Black Hole in M32 from HST/FOS Spectra. II. Axisymmetric Dynamical Models*, ApJ, 493, 613 8

van der Marel R. P., Franx M., 1993, *A new method for the identification of non-Gaussian line profiles in elliptical galaxies*, ApJ, 407, 525 60, 63, 166

van der Marel R. P., Rix H. W., Carter D., Franx M., White S. D. M., de Zeeuw T., 1994, *Velocity Profiles of Galaxies with Claimed Black-Holes - Part One - Observations of M31, M32, NGC 3115 and NGC 4594*, MNRAS, 268, 521 63

van der Marel R. P., van den Bosch F. C., 1998, *Evidence for a $3 \times 10^8 \, M_\odot$ Black Hole in NGC 7052 from Hubble Space Telescope Observations of the Nuclear Gas Disk*, AJ, 116, 2220 15

Vega Beltrán J. C., Pizzella A., Corsini E. M., Funes J. G., Zeilinger W. W., Beckman J. E., Bertola F., 2001, *Kinematic properties of gas and stars in 20 disc galaxies*, A&A, 374, 394 168, 169, 197

Verolme E. K., et al., 2002, *A SAURON study of M32: measuring the intrinsic flattening and the central black hole mass*, MNRAS, 335, 517 15, 90, 103

Volonteri M., Haardt F., Madau P., 2003, *The Assembly and Merging History of Supermassive Black Holes in Hierarchical Models of Galaxy Formation*, ApJ, 582, 559 24

Volonteri M., Lodato G., Natarajan P., 2008, *The evolution of massive black hole seeds*, MNRAS, 383, 1079 24

Weitzel L., Krabbe A., Kroker H., Thatte N., Tacconi-Garman L. E., Cameron M., Genzel R., 1996, *3D: The next generation near-infrared imaging spectrometer.*, A&AS, 119, 531 130

Whitmore B. C., Schechter P. L., Kirshner R. P., 1979, *Velocity dispersions in the bulges of spiral galaxies*, ApJ, 234, 68 197

Acknowledgements

Winsall M. L., Freeman K. C., 1993, *Velocity distributions in spherical elliptical galaxies. II - Measuring line-of-sight stellar velocity distributions*, A&A, 268, 443 63

Wold M., Lacy M., Käufl H. U., Siebenmorgen R., 2006, *The nuclear regions of NGC 7582 from [Ne II] spectroscopy at 12.8 μm - an estimate of the black hole mass*, A&A, 460, 449 15

Woo J.-H., Treu T., Malkan M. A., Blandford R. D., 2008, *Cosmic Evolution of Black Holes and Spheroids. III. The M_{BH}-σ_* Relation in the Last Six Billion Years*, ApJ, 681, 925 30, 31

Wozniak H., Combes F., Emsellem E., Friedli D., 2003, *Numerical simulations of central stellar velocity dispersion drops in disc galaxies*, A&A, 409, 469 118, 166

Wright C. O., Egan M. P., Kraemer K. E., Price S. D., 2003, *The Tycho-2 Spectral Type Catalog*, AJ, 125, 359 116

Wyithe J. S. B., 2006, *A log-quadratic relation between the nuclear black hole masses and velocity dispersions of galaxies*, MNRAS, 365, 1082 26

York D. G., et al., 2000, *The Sloan Digital Sky Survey: Technical Summary*, AJ, 120, 1579 25, 149

Younger J. D., Hopkins P. F., Cox T. J., Hernquist L., 2008, *The Self-Regulated Growth of Supermassive Black Holes*, ApJ, 686, 815 6, 22, 106

Yu Q., Tremaine S., 2002, *Observational constraints on growth of massive black holes*, MNRAS, 335, 965 5

Zel'dovich Y. B., 1964, *The Fate of a Star and the Evolution of Gravitational Energy Upon Accretion*, Soviet Physics Doklady, 9, 195 2

Acknowledgements

Foremost I would like to thank my thesis advisor Ralf Bender for giving me the opportunity to explore such an interesting field of modern astrophysics, for investing so much observing time in this project and for giving me the time and support I needed. I also benefitted much from the opportunity to present my work at several international conferences.

Very special thanks go to my thesis supervisor Roberto Saglia for his guidance, support and endless patience during all those years.

This thesis would not be what it is without Jens Thomas. Only thanks to his expertise and his continuous support and help particularly in the modelling part I was able to derive black hole masses from my data. I am also indebted to Peter Erwin and his detailed knowledge about photometry and pseudobulges. He also contributed a lot to this work. Many thanks also to Karl Gebhardt, who provided and showed me the MPL code and who was available for discussions about kinematics and everything else during the times he spent at MPE.

I would like to thank the late Peter Schuecker, who supervised my Diploma thesis and, as a member of my Ph.D. committee, was a valuable contact person and advisor during the first part of my Ph.D. thesis.

Further sincere thanks go to the members of the infrared group at MPE, in particular to Ric Davies for his support during the observations and for patiently answering all the questions I had about SINFONI, data reduction, the laser guide star and many other things, to Frank Eisenhauer for introducing me to SINFONI data and data reduction and to Stefan Gillessen for his support during the observations, especially during the first observing run when I was still very inexperienced.

I wish to thank Karina Kjaer and Mariya Lyubenova for general discussions about IFU data reduction and line index measurements.

Acknowledgements

Furthermore I would like to thank Alexis Finoguenov who provided relaxation and distraction at MPE with his excellent salsa classes.

I am very grateful to my father, who once raised my interest in physics and astronomy and who always supported me in whatever I was doing.

Very special thanks go to Robert Wagner not only for scientific discussions and proofreading the manuscript, but also for being always there when I needed him and for supporting and encouraging me. I would also like to thank his parents Gabriele and Peter for making me feel at home and for always supporting and helping me whenever I had problems.

Finally I would like to thank my friends Nadia Matter, Rita Przewodnik, Yu-Ying Zhang and Yuliana Goranova for reminding me that there is a life beyond the thesis and for not being angry despite the numerous long periods when I did not have time for them.

Die VDM Verlagsservicegesellschaft sucht für wissenschaftliche Verlage abgeschlossene und herausragende

Dissertationen, Habilitationen, Diplomarbeiten, Master Theses, Magisterarbeiten usw.

für die kostenlose Publikation als Fachbuch.

Sie verfügen über eine Arbeit, die hohen inhaltlichen und formalen Ansprüchen genügt, und haben Interesse an einer honorarvergüteten Publikation?

Dann senden Sie bitte erste Informationen über sich und Ihre Arbeit per Email an *info@vdm-vsg.de*.

Sie erhalten kurzfristig unser Feedback!

VDM Verlagsservicegesellschaft mbH
Dudweiler Landstr. 99　　　　　　Telefon　+49 681 3720 174
D - 66123 Saarbrücken　　　　　　Fax　　　+49 681 3720 1749
www.vdm-vsg.de

Die VDM Verlagsservicegesellschaft mbH vertritt

Printed by Books on Demand GmbH, Norderstedt / Germany